CLIMATE CHANGE, MIGRATION AND CONFLICT IN BANGLADESH

This book explores the relationship between climate change–induced migration and conflict in Bangladesh – one of the most ecologically fragile countries in the world. It explores why people migrate from their original place of land and how the migration of people with a different background to an ethnically distinctive region due to environmental changes can become a source of conflict and violence between the host peoples and migrants. The volume focuses on the Chittagong Hill Tracts (CHT), which has experienced long-standing ethnopolitical conflict due to the settlement and migration of the Bengali people from the plain land of Bangladesh. This settlement and migration were mainly caused climatic events such as floods, cyclones, sea-level rise, and disasters. It traces the history of the ethnic conflict in the region and presents key findings from the field, as well as the dynamics of everyday politics in the region. This volume also highlights how internally climate-displaced people generate violence and civil strife in the major urban cities through their settlements in slums.

The volume will be of great interest to scholars and researchers of environmental studies, human geography, migration and diaspora studies, public policy, social anthropology, and South Asian studies.

Md Rafiqul Islam is a professor at the Department of Peace and Conflict Studies, University of Dhaka, Bangladesh. Dr. Islam has a PhD from Flinders University, Australia, and his thesis was awarded the John Lewis Silver Medal in South Australia. He had previously completed his bachelor's degree in political science and has a master's and an MPhil in peace and conflict studies from the University of Dhaka, Bangladesh. He has also completed an MA degree in environmental security and peace from the University for Peace, Costa Rica. His primary research interests include climate change, migration, refugees, peace, conflict, and development. Dr. Islam also has a keen interest in global politics, security, and peace issues and has published 40 articles and chapters across several publication based globally.

Climate Change, Migration and Conflict in Bangladesh

Md Rafiqul Islam

LONDON AND NEW YORK

First published 2024
by Routledge
4 Park Square, Milton Park, Abingdon, Oxon OX14 4RN

and by Routledge
605 Third Avenue, New York, NY 10158

Routledge is an imprint of the Taylor & Francis Group, an informa business

British Library Cataloguing-in-Publication Data
A catalogue record for this book is available from the British Library

ISBN: 978-1-032-21563-1 (hbk)
ISBN: 978-1-032-55427-3 (pbk)
ISBN: 978-1-003-43062-9 (ebk)

DOI: 10.4324/9781003430629

Typeset in Sabon
by SPi Technologies India Pvt Ltd (Straive)

Contents

Illustrations

Figures

Tables

Preface

The link between climate change and migration already has an established literature. Empirical studies on the topic point out to instances of people and communities being displaced because of climate change, as they lose their livelihoods and often their homes due to the adverse effects of climatic events. For these people and communities, displacement due to climate change can be temporary at first, but in time they may find their situation unchanging. This can happen due to their livelihood and financial conditions not improving in the long run, or not receiving adequate support from the government that they might sorely need. This leads to them migrating to other areas in search of a renewed, stable life. These people are referred to as climate migrants, indicating the reasons behind their search for new opportunities in otherwise better environments. The literature on this topic is extensive with a theoretical foundation and a plethora of empirical research. However, when exploring theoretical discussion on whether or not climate change has any influence in creating conflict, the literature seemed to be woefully inadequate.

On this note, I have been influenced to write this book by reading a particular book named *Climate and human migration: Past experiences, future challenges* by Robert McLeman. The book explores how climate change and migration are interlinked, bringing attention to the scale of the influence the phenomenon really has on the lives of millions of people. At the concluding sections of the book, the author pointed out the lack of a solid theoretical basis and an empirical foundation on the relationship between climate change and migration, despite climate change having definite potential of creating conflicts in societies just from its contributions in creating so many climate migrants all across the world. Moreover, works by Nil Peter Gleditsch, Idean Salehyan, and Rafael Reuveny on climate change and conflict relationships also inspired me to see how climate change, migration, and conflict interplay in a highly climate affected country like Bangladesh.

The works of the previous authors mentioned the need for further research and empirical studies on how climate change and the resulting migration can lead to conflict. Furthermore, they indicated how studies should be carried out in countries that are most affected by climate change to explore this relationship between climate change–induced migration and conflict. This struck me as very relevant for the contemporary era, where the effects of climate change are becoming more and more severe, and disproportionately so for developing and underdeveloped countries. News reports and periodicals have periodically mentioned particular areas of climate-affected countries experiencing conflict. The reported conflicts, as these sources mention, often stem from an influx of migrants settling in an area for new and better opportunities which they have lost at their previous homes, with some losing their homes entirely. The conflicts have always been on smaller scales, leading to at most civil strife and local unrest, and remaining confined to a particular area.

The conflicts resulting from climate migrants have never been on the scale of conflicts between two nations, which may be the cause of the apparent lack of published studies on the topic, if any. This suggests a clear indication that a link exists between migration induced by climate change and conflict. A theoretical discussion on this topic was difficult to identify, and I was unable to find any foundational work or empirical study that focused on it. I then recalled McLeman's comment on how there should be more research on the topic and the need for empirical studies on countries most affected by climate change. This led me to a line of inquiry that hinted toward a relevant case study on this very topic. Of the countries most affected by climate change, Bangladesh, a South Asian country of the coast of the Bay of Bengal, is relevant in this regard. The country has continued to have conflicts ongoing for decades, as the people in the Chittagong Hill Tracts (CHT) are struggling for their own rights to self-determination and self-rule with the people from the low-lying plain lands. Additionally, this particular country is among those that are highly affected by climate change. The country experiences cyclones, floods, and river erosions on an increasing scale and intensity each year, and migrations resulting from these events are frequent within its borders. People often lose their livelihoods and even their homes when a cyclone and river erosion crumbles their lands away or make them unhospitable, leading the people to migrate for better opportunities.

The conflicts taking place in the CHT region are largely due to how the demographics have shifted because of migrants from the plain lands moving to this hilly region of the country. In the mainstream literature of the CHT issue, these people are called the Bengali settlers. The successive military governments after 1975 settled these Bengali people to make demographic balance with the local people in the CHT. The migrants have disturbed the particular communal way of living and land ownership of the ethnic communities in the region, and the resulting conflicts have ballooned to the present scale over decades of civil unrest. The demographic of the region shifted from a mere 7 percent of migrants during 1971, to near 50 percent in 2020. With regards to the migrants arriving in larger numbers in the area that have contributed significantly to the inter-communal conflicts there, some important findings have surfaced. A large percentage of the migrants in the region have not settled there due to reasons such as lucrative financial incentives or government sponsorship, but due to climatic events ruining their existing livelihoods. These migrants viewed the promise of a stable settlement and livelihood in the CHT region as an opportunity that was difficult to avoid taking, thus bringing the issue of climate change–induced migration deep into this mountainous region of Bangladesh. This line of inquiry thus has led me to explore this relationship between climate change–induced migration and conflict, using Bangladesh as a relevant case study. This work is but one step of a series of steps in exploring this link between these concepts in an empirical setting.

This book highlights that climate changes have profound impacts on displacing people from their place of land. The sudden and slow-onset processes of climatic events complicate the living conditions and livelihood facilities of the poor people who eventually migrate short distances, sometimes required as part of a basic survival strategy, and sometimes for diversifying their income. The climate affected people only migrate to the urban areas and the CHT region in Bangladesh as and when they lose everything, and nothing remains left for them to support their lives in their original homes. Sudden floods, cyclones, and riverbank erosions lead to such drastic and dire conditions when the poor people in the rural and coastal regions are forced to migrate long distances. Migration is sometimes considered and seen as the adaption strategy for the poor people who just were not able to survive and manage their livelihoods with existing conditions. However, migration of these climate-affected people sometimes become a burden for the host society and thus acts as part of a set of triggering factors for conflict and violence in host societies. This relationship has been explored in the urban cities and the CHT region in Bangladesh. In the urban cities such as Dhaka, the migration and conflict relationship is less perceived than the CHT. Bengali settlement and migration in the post 1975 period complicated the entire socioeconomic conditions through changing the demographic composition in the region. The demographic transformation eventually equalizes the ratio of the Bengali and tribal people living in the mountainous area. At the same time, increasing population put enormous pressure on the existing resources such as land, forest, and social positions and has generated resource scarcity in the process.

Competition over natural resources and social positions has resulted in events involving resource capture, plundering, discrimination, and marginalization of the indigenous tribal people. The indigenous tribal people perceive the land and resources as their own and common property as part of their communal ownership system, as they have enjoyed these resources over many generations as such. On the other hand, the migrant people started to perceive the land and resources of the CHT as their own after their settlement there. These two competing perceptions are the direct outcome of a large-scale migration of the Bengali people in the region, which has been the source of violence and conflict.

This book contributes to this important topic of climate change, migration, and conflict in Bangladesh. It hints that migration of climate change-affected people may not be an adaptation option in a place where people with different backgrounds live. Migration and settlement of a group of people in a place where an ethnic minority community live may instead generate long-standing social conflicts. Such situations can also force the minority people to become marginalized and sometimes be seen as part of a systematic process of exclusion. Conflicts resulting from such migration can also be perceived in the urban areas where the government failed to manage the climate migrant people with basic needs and facilities. The struggles of these migrants that take place as a direct result of such management failures can spiral into other unavoidable yet unfortunate outcomes that in the end stem from the basic drive to survive and prosper. It is felt that this book can shed some much needed light on the plight of migrants, who travel to places for better wages and are in the end helpless against the ruthless force of nature. The interplay among climate change, migration, and conflict is the highlight of this book. It is felt that through the contextual focus of such an interplay in a country such as Bangladesh that experiences the effects of climate change first-hand that this book may invite some much-needed discourse. It is also strongly hoped that this discourse can lead to a definite recourse to effective action in the very near future.

Acknowledgments

This book is the outcome of my PhD thesis that I conducted from 2015 to 2019. Thus, first of all, I acknowledge the contribution, scholarship, field trip support, and logistics provided by Flinders University, Australia, by which I completed my PhD studies. I duly acknowledge the contribution of my principal supervisor Professor Susanne Schech and associate supervisor Dr. Udoy Saikia, School of Arts, Humanities and Social Sciences, Flinders University. I am truly grateful to the Australian government scholarship that enabled me to stay in Australia and finish my PhD studies.

I am grateful to my parents who had sent me to school to learn something. My parents did not get scope to go to school, but they ensured all facilities for us to study for seeing the world. This is how most of the parents in the rural area of Bangladesh have sacrificed for their children.

I acknowledge the support of my research associate Muhammad Mazedul Haque who thoroughly checked the referencing style of this manuscript. Haque gave his effort to read this manuscript and provide his valuable suggestion.

I am grateful to my wife Taslima Islam and daughter Anushka Awrin for their mental support and encouragement for doing good deeds in my life. I also acknowledge the support of my brothers and sisters for their good wishes.

Last but not least I acknowledge the contribution of my teachers from the primary school to the university who have taught and encouraged me to become a true teacher.

1 Introduction

Contextualizing the study

Introduction

This book addresses a topic of particular concern to researchers in the social sciences, namely whether and how climate change–induced migration causes conflict and violence. It explores these connections through a local level case study to analyze the impacts of climatic events on decisions of the migration of people in Bangladesh, as well as the implications of the migration for social, economic, and political change in the urban areas and other areas where the poor and climate change induced people migrate and settle – in particular in the Chittagong Hill Tracts (CHT) in Bangladesh. This is to note that Chittagong Hill Tracts is a hilly area where the minority ethnic indigenous people[1] live and the area experienced violent conflict until 1997 (Adnan, 2004; Chakma, 2010; Mohsin, 1997). Moreover, a small-scale social conflict is still continuing in the area between the local people and the migrated Bengali people (Choudhury, Islam, & Alam, 2017). Moreover, the urban areas such as Dhaka and Chittagong absorb a large number of climate change–induced internal migrants every year (Ahsan, Kellett, & Karuppannan, 2016; Luetz, 2018) who put enormous pressure on the existing resources and services. This pressure generates scarcity that often triggers civil violence and unrest in the urban areas. This book explores this connection of climate change, migration, and conflict, as well as provides an explanation of the process of migration and conflict in Bangladesh. Bangladesh is one of the most climate affected countries in the world (Roy, 2011). The worst affected people are left with no option other than to migrate to cities and any suitable place including the CHT for livelihood and shelter. The CHT is a conflict-prone region in Bangladesh where it is presupposed that Bengali settlement and migration have played a significant role in causing and escalating the conflict. Based on the theories of climate change, migration, and conflict, this book explores the relationship between climate change and the migration decisions, and migration and conflict taking a local level case study.

This book is written based on qualitative methods. The information and data have been collected through interviews among experts and key informants. Interviews were conducted among the key informants in the urban

DOI: 10.4324/9781003430629-1

slums and the CHT region. The key informants are climate migrants who have been displaced for climatic events. First, interviews were collected from ten climate migrants who live in the slums in Dhaka city to understand their situation in the place of origin and how they have migrated to the urban areas. Then, interviews with Bengali climate migrants in the CHT were also conducted in order to know how they have migrated to the CHT and how climatic events affected them in their place of origin. The author conducted the interviews of total 20 self-claimed climate affected people (Appendix 1), referred to as climate migrants (CMs), about their lived experience of the climatic events they faced, their actions in response to these, the process of migrating to the urban areas and the CHT, and engaging with its people, society, and politics. The 20 CMs were randomly selected from those who confirmed that they had migrated as a result of climate change events at their place of origin and gave their consent to take part in a follow-up interview. The author also interviewed 15 experts who hold strategic positions in academia, the nongovernment sector, political leaders, and activists (Appendix 2). Participants for the interviews with experts were selected based on their knowledge and expertise on climate change, migration, and/or the CHT conflict. However, it should be remembered that the experts have imperfect knowledge on CHT affairs, in particular, on the ways in which the migration decisions of Bengali people in the CHT have been affected by climatic events.

The impacts of climate change on conflict came to light in the early 21st century through publications and media reports concerning the security implications of such change (Barnett, 2003; Barnett & Adger, 2007). Some researchers and research organizations outlined the effects of climate change on social disturbance, political instability, and conflict in the national and international context (Rahman, 1999; Schwartz & Randall, 2003; WBGU, 2008). The former secretary general of the United Nations, Ban Ki Moon, claimed that climate change played a role in the conflict and humanitarian crisis in Darfur, Sudan, and that "similar types of environmental problems are equally causing violence in other African countries" (Nordås & Gleditsch, 2007: 629). The recent conflict in Syria has also been partly attributed to the prolonged droughts and their impact on economic conditions (Gleick, 2014). In its fifth assessment report the Intergovernmental Panel for Climate Change (IPCC), the main agency providing scientific information about climate change issues, argues that climate change will endanger the human security of millions of people across the world, and this will be one of the primary sources of insecurity, political instability, and civil conflicts in many countries (Adger et al., 2014: 777). The IPCC report also postulates

> Climate change threatens human security because it undermines livelihoods, compromises culture and individual identity, increases migration that people would rather have avoided, and because it can

> undermine the ability of states to provide the conditions necessary for human security. Changes in climate may influence some or all of the factors at the same time. Situations of acute insecurity, such as famine, conflict, and socio-political instability, almost always emerge from the interaction of multiple factors.
>
> (Adger et al., 2014: 762)

Over the past two decades, research into the relationship between climate change and conflict has produced many publications and reports, and there is now broad public awareness of the security implications and potential for conflict resulting from climate change (Barnett & Adger, 2007; Hendrix & Salehyan, 2012; Reuveny, 2007; Theisen, Holtermann, & Buhaug, 2012).

Despite growing scholarly and public awareness, there is a lack of consensus among researchers, policy analysts, and world leaders on the relationship between climate change and conflict (Mach et al., 2019; Nordås & Gleditsch, 2015). The main issue is that climate change is not the sole source of conflict and violence. Other factors, such as social, economic, and political issues, influence the links between climate change and conflict (Gleditsch, 2012; Mach et al., 2019; Salehyan, 2014; Selby & Hoffmann, 2014). While resource scarcity due to climate change and competition between and among groups for resources can be key factors in conflict and violence, this type of conflict is more prevalent in economically poor countries (Evans, 2011; Homer-Dixon, Boutwell, & Rathjens 2011), which suggests that poverty is also a root cause. Climate change–induced migration is another potential source that may create or trigger further conflict when migrants complicate social, economic, and political conditions in the receiving societies (Brzoska & Fröhlich, 2016; Gleditsch, Nordås, & Salehyan, 2007; Raleigh, Jordan, & Salehyan, 2008; Reuveny, 2007, 2008). Indeed, among the possible connections between climate change and conflict, climate change–induced migration is the most influential and far-reaching in climate hotspots because climate change–induced migration affects the resource base, cultural issues, and the social and political systems of the host society.

To date, the cases that have been reported as climate change–induced conflict are subject to criticisms due to lack of reliability of information, limited validity of the research results, and methodological weaknesses in exploring the relationships between climate change and conflict (Gleditsch & Nordås, 2014; Nordås & Gleditsch, 2015; Salehyan, 2014; Theisen, Gleditsch, & Buhaug, 2013). Thus, researchers have proposed that more empirical work should be conducted based on local level case studies in climate hotspots to substantiate the assumed links between climate change and migration; and between climate change and conflict (McLeman, 2014). For example, Hendrix suggested that "the most climate exposed countries such as Bangladesh and Haiti" should receive more attention from researchers interested in the relationship between climate change and conflict (Hendrix, 2018:

190). This book endeavours to contribute to this avenue of knowledge by exploring whether and how climate change has contributed to migration and conflict in the urban areas and CHT in Bangladesh. Understanding this connection can assist with addressing the factors that fuel the current conflict and finding new peacebuilding options in the region. This overarching question focuses on three issues: (1) To what extent climatic events played a role in the migration of people in Bangladesh? (2) Where do the worst affected people normally go and settle? (3) How have migration and settlement of the CMs contributed to the host places, such as urban areas and the CHT region? By examining the ways in which climate change, migration, and conflict are interconnected in the local level in Bangladesh, this book aims to provide a broader understanding of conflict in climate change–affected countries and identify lessons that can be learned for managing the predicted increase in climate change–induced migration.

Why climate change, migration, and conflict research?

The study of climate change, migration, and conflict is timely as scientific research has already established that climate change is a real phenomenon and increasing global warming will alter the environment, sea level, and livelihood options for human beings (IPCC, 2014b). Climate change in this regard refers to the changing pattern of weather conditions in the global and regional context. It is the fluctuation of temperature, extreme heat or cold, that greatly impacts on human beings. The IPCC (2007) defines climate change as "changes in the state of the climate that can be identified (for example, using statistical tests) by changes in the mean and/or the variability of its properties, and that persists for an extended period, typically decades or longer. It refers to any change in climate over time, whether due to natural variability or as a result of human activity". The United Nations Framework Convention on Climate Change (UNFCCC) defines climate change as:

> Climate change means a change of climate which is attributed directly or indirectly to human activity that alters the composition of the global atmosphere and which is in addition to natural climate variability observed over comparable time periods.
>
> (UNFCCC, 1992, article 1)

Climatic events such as global warming, sea-level rise, drought, floods, and disasters are occurring more frequently than before with adverse impacts on the environment and human beings across the world (IPCC, 2014a; Field, 2014; Hinkel et al., 2014; Pittock, 2017; Romm, 2018; Walsh et al., 2016). The IPCC report of 2014 states that the last three decades have experienced higher earth surface temperatures than any preceding decades since 1980 (IPCC, 2014: 40). The scientific evidence demonstrates that climate change is occurring on a global scale. This change is natural and human-induced (i.e., man-made). The following facts and figures present the scenario of climate change and its impacts on the earth and on human beings.

The global land and ocean surface temperature increased significantly from 1950 to 2020 in comparison to the decades before 1950 (Figure 1.1). The fifth assessment report published by the United Nations Scientific Panel on Climate Change states "the atmosphere will warm up by 1.5 degrees (Celsius) by 2040 if the current rate of greenhouse emission continues" (Davenport, 2018).

This warming will lead to inundation of the coastline and intensification of droughts and poverty" (Davenport 2018). Global warming has had its effects on sea-level, which shows a rising trend (Figure 1.2). Sea-level is increasing across the world for several reasons, but the primary cause is the anthropogenic global warming that leads to melting the land-based ice and glaciers.

The IPCC sixth report (2021: SPM 6) outlines that

> Global mean sea level increased by 0.20 [0.15 to 0.25] m between 1901 and 2018. The average rate of sea level rise was 1.3 [0.6 to 2.1] mm yr^{-1} between 1901 and 1971, increasing to 1.9 [0.8 to 2.9] mm yr^{-1} between 1971 and 2006, and further increasing to 3.7 [3.2 to 4.2] mm yr^{-1} between 2006 and 2018 (high confidence). Human influence was very likely the main driver of these increases since at least 1971.

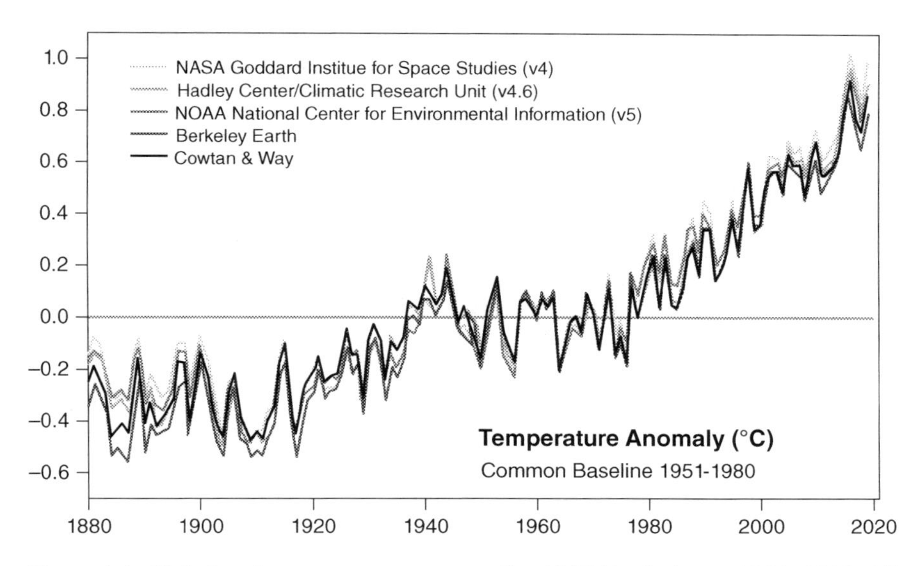

Figure 1.1 Global surface temperature anomalies (GSTA) relative to 1850–1900. All the sources predict the increasing of temperature in the globe

Source: IPCC (2021).

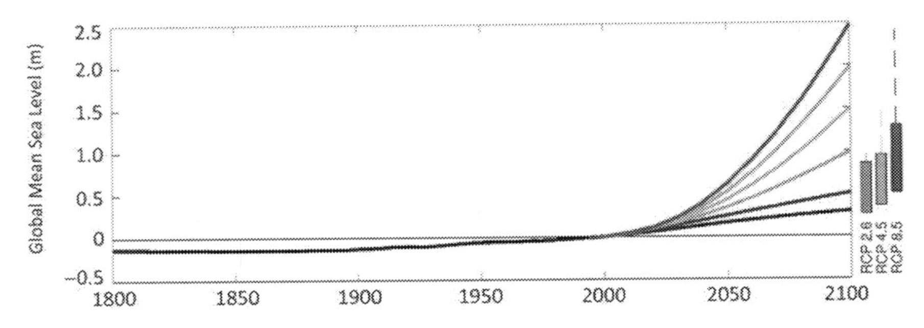

Figure 1.2 Global mean sea-level change

Source: IPCC (2021). Copyright: European Environmental Agency. https://www.eea.europa.eu/data-and-maps/figures/projected-change-of-global-mean.

The big concern for the sea-level rise is the inundation of some regions situated at or close to sea level (Connell, 2003). For example, some low-lying countries such as the Maldives, Tuvalu, and Bangladesh are highly exposed to sea-level rise. The melting of ice and decreasing rate of ice levels cause this alarming concern. The IPCC sixth assessment report (2021: SPM9) outlines that "In 2011–2020, annual average Arctic sea ice area reached its lowest level since at least 1850 (high confidence)."

The pace of greenhouse gas concentration has increased from the 1950s, and after 2000 the trend has been increasing more rapidly due to human activities, such as burning fossil fuels and land-use changes. From 2000 to 2010 greenhouse gas emissions were the highest in history (IPCC, 2014: 44) (Figure 1.1). The sixth assessment report (2021: SPM 9) provides that "In 2019, atmospheric CO2 concentrations were higher than at any time in at least 2 million years (high confidence), and concentrations of CH4 and N2O were higher than at any time in at least 800,000 years (very high confidence)." Human activities and industrialisation in developed and developing countries are causing anthropogenic CO_2 emissions, which are inflicting detrimental impacts on human and natural systems.

The IPCC sixth assessment report (2021) highlights that the increasing climate changes have contributed to increases in the frequency of cyclones, prolonging of droughts and wild fires in many places across the world. The frequency of concurrent heatwaves and droughts vary in the world due to the global warming and climatic changes. Global warming in particular contribues to "increases in the frequency and intensity of hot extremes, marine heatwaves, and heavy precipitation, agricultural and ecological droughts in some regions, and proportions of intense tropical cyclones, as well as reductions in Arctic sea ice, snow cover and permafrost" (IPCC, 2021 SPM19) (Figure 1.3).

The IPCC fifth assessment report highlighted these key issues for sensitizing the international community about the human interventions to climate change. The IPCC Summary Report 2014 stated:

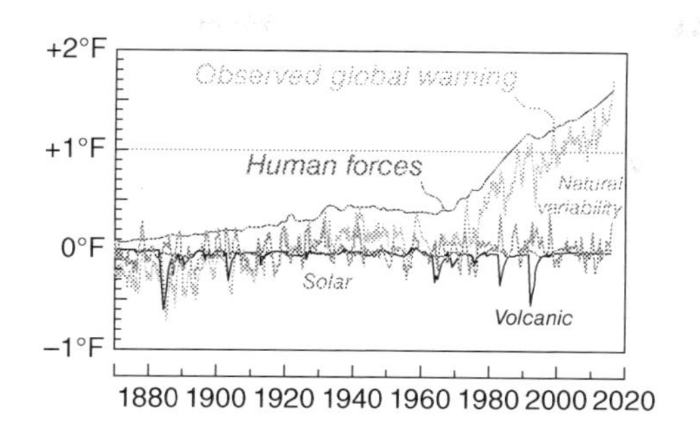

Figure 1.3 Forces affecting global temperature

Source: IPCC (2021).

> In recent decades, changes in climate have caused impacts on natural and human systems on all continents and across the oceans. Impacts are due to observed climate change, irrespective of its cause, indicating the sensitivity of natural and human systems to changing climate.
>
> (Pachauri et al., 2014: 6)

Changes in climate are also causing droughts, floods, cyclones, and coastal erosion across the world. The new IPCC report details that, depending on the region, climate change has already worsened extreme heat, drought, fires, floods, and hurricanes, and these will only become more damaging and destructive as temperatures continue to rise. The IPCC's 2018 "1.5°C Report" had entailed the differences in climate consequences in a 2°C versus 1.5°C world, as summarized by Bruce Lieberman (IPCC, 2021). The low-lying regions in Asia and drought-prone areas in Africa are highly affected by such changes. Many people living in these areas are poor, which makes them more vulnerable (Field, 2014; Hunter, 2005), as "climate change will amplify existing risks and create new risks for natural and human systems" (Pachauri et al., 2014, 62). For example, researchers predict that increasing climate change will prolong droughts, cause extreme cold and heat that will affect human health, result in disease, reduce agricultural production, increase costs of agriculture and decrease the ability of people to live in their homes (Hendrix & Salehyan, 2012; Knox et al., 2012; Theisen, Holtermann, & Buhaug 2012). Climate change is also destroying resources, decreasing resource stocks, and reducing the livelihood options on which human beings depend (Afifi & Warner, 2008; IPCC, 2014b; Bellard et al., 2012; McMichael & Lindgren, 2011; Stern, 2006; Urry, 2015).

But the question is, what do people do in the face of climatic events? Climate change affected people respond in different ways, such as adapting to climate change or abandoning their homes (Reuveny, 2008; Hugo, 2013). People who leave their homes are forced to migrate, or decide to migrate voluntarily. They may do so temporarily or permanently, within the country or across national borders. They migrate to avoid the vulnerabilities associated with climate change, for example, to find shelter and to improve or diversify their income (Baldwin, 2017; Hugo, 2013; McLeman, 2014). Here, displacement refers to a situation when climatic events indiscriminately force people to leave their place of origin. According to Stapleton and colleagues (2017: 6), "displacement refers most commonly to instances where there is no choice but to move, either temporarily or permanently, within or across borders." Conversely, migration denotes the process of people moving from one place to another temporarily or permanently (McLeman, 2014). This process of moving is determined either by choice or by force as a result of natural disasters and political problems.

The people displaced due to climate change are termed variously as *climate refugees* or *climate migrants*. Climate refugees have been defined as "those people who migrate across borders for the direct effects of climate change, such as sea-level rise, extreme weather events, droughts and water scarcity" (Biermann & Boas, 2012: 18–19). According to this definition, people who have been internally displaced due to climate change are not recognised as refugees. Harvard legal scholars Docherty and Giannini similarly define a climate change refugee as:

> An individual who is forced to flee his or her home and to relocate temporarily or permanently across a national boundary as the result of sudden or gradual environmental disruption that is consistent with climate change and to which humans more likely than not contributed.
>
> (Docherty & Giannini, 2009: 361)

Thus, people who are forced to move by climatic events are only recognised as refugees when they cross an international border. Although climate change–induced migrants are labelled in some research and policy documents as refugees (Myers, 1993, 2002), the United Nations High Commission for Refugees (UNHCR) and some researchers have criticized the term *climate refugees*. They argue that refugee status is only given to people who are displaced by political emergencies, conflicts, persecution, and other factors recognized under the Geneva Convention and related protocols and who have crossed international borders temporarily or permanently (McAdam, 2010, 2012). Therefore, the term *climate migrants* has been proposed to define people displaced by climate change events, such as increasing temperature, floods, sea-level rise, drought, cyclones, and coastal erosion (Laczko & Aghazarm 2009) and moved to another location either as international or internal migrants (McLeman & Smit, 2006). This book uses the term *climate migrant* to refer to people who have attributed their decision to migrate at least in part to climatic events, such as floods, sea-level rise, cyclone, drought, and coastal erosion.

Whichever label is given to climate displaced people, they are now a great concern for the affected regions, as well as the potential migrant–receiving

Table 1.1 Estimates of climate displaced people and future predictions

Already displaced people	*Future predictions*
1 Around 24 million people have been displaced by floods, famine, and other environmental disasters (UNHCR, 2002: 12)	1 Around 162 million people would be at risk and displaced due to sea-level rise by 2050 (Myers, 2002, cited by Adamo, 2009: 18; also see Biermann & Boas, 2010: 68)
2 In the last seven years, climate or weather-related disasters have displaced 22.5 million each year (IDMC, 2015: 8)	2 Around 250 million people would be refugees from climatic events by 2050 (Christian Aid, 2007, cited in Biermann & Boas, 2010: 68)
3 In 2014, more than 19.3 million people were displaced by disasters in 100 countries (IDMC, 2015: 8)	3 Climatic events by 2050 would permanently displace 200 million people (Stern, 2006: 56)
4 50 million people were displaced by environmental causes by 2010 (UNFCCC, 2007: 42)	4 Around 140 million people could be displaced by 2050 in sub-Saharan Africa, South Asia, and Latin America (World Bank, 2018: XIV)

Source: Compiled by the author from different sources.

regions and nations. Due to the growing speed and impact of climate change, more people are being displaced and migrating from their homes. In order to alert the international community, various studies have provided estimates of existing human displacement and migration from climate change as well as scenarios for the future projected migration (Table 1.1).

While these figures indicate that climatic events have already displaced hundreds of millions of people in recent years, the future predictions are alarming, particularly for developing regions. Most of the poor and underdeveloped countries have limited infrastructure and technical knowledge, as well as depending on a subsistence economy for livelihood. Due to a lack of modern technology, knowledge, and economic capability, climate affected people in underdeveloped regions are unable to manage climate vulnerability and thus are forced into displacement and migration. Countries in Asia, Latin America, and Africa are more exposed to climate change–induced displacement and migration. The climate change–induced internal migration rate is much higher than the international migration rate because climate affected people usually attempt to find shelter and livelihood options within rather than outside their countries (Faist, 2000; Hugo, 2013; Warner, 2010). Currently, climate change hazards are causing internal displacement in low-land areas across the world. For example, research by the Internal Displacement Monitoring Centre (IDMC) states that "between 2008 and 2016, sudden-onset events were responsible for 99 percent of internal displacement: an average of 21 million people annually" (IDMC, 2017: 4–5). Werz and Conley (2012: 3), citing the United Nations *Human Development Report 2012*, stated that "worldwide, there are already an estimated 700 million internal migrants—those leaving their homes within their own countries—a number that includes people whose migration is related to climate change and environmental factors."

The displaced and migrated people can have detrimental impacts on the social, economic, and political context in the affected countries (Gemenne et al., 2014; Matthew et al., 2010; Trombetta, 2014). Their migration has implications for security and political stability in national and international contexts. For example, researchers and security analysts argue that climate change will complicate individual, national, and international security through scarcity of vital resources, such as water (Adger, 2010; Barnett, 2003; Brown, Hammill, & McLeman 2007; Scheffran & Battaglini, 2011). Some researchers have redefined the human security issue by linking it with the trajectory of climate change (Barnett & Adger 2007). In its fifth assessment report, the IPCC proposes that climate change has the impact of shrinking the livelihood options for people, which in turn exacerbates human insecurity and conflict (Adger et al., 2014). As a result, the ways in which climatic events create or impact on conflict has been the subject of a growing number of research studies.

This raises a further question –what is climate change conflict? To date, there is no separate definition developed to describe climate change conflict. Researchers have defined small-scale incidents of violence, civil strife, riot, killing, insurgencies, communal conflict, and ethnic war as climate change conflict if these incidents occur in climate hotspots. The conflict and hostilities between countries over water and other resources in climate affected regions are also referred to as climate change conflict. Ide and Scheffran have attempted to define conflict and violence originating from climate change issues as "(a) small-scale conflicts having the 25 deaths threshold, (b) concern the use of natural resources (rather than government and territory), and (c) take place between the social groups" (Ide & Scheffran, 2014: 269). Indeed, climate change conflict is not a new form of conflict but an annexation of terminology in the lexicon of conflict analysis. It mainly occurs between and among groups within a country due to resource scarcity and resource sharing.

The definition of climate change conflict is closely connected with the conventional definition of conflict. In their analysis of the climate-conflict situation in African countries, Hendrix and Salehyan (2012: 35) use the term *social conflict* to describe internal conflicts such as demonstrations, riots, strikes, communal conflict, and antigovernment violence that have originated primarily from climate change–induced adverse effects. In the conflict literature, the concepts of internal conflict and social conflict are used interchangeably and sometimes refer to the same things while analyzing conflict in the internal context of a state. In this sense, the definition of climate change conflict is not distinctive and does not represent a separate conflict issue. Burke et al. (2015) define conflict based on interdisciplinary perspectives, giving a long list of the conflicts that constitute climate change conflict. For these writers, climate change conflict represents

> many types of human conflict including both interpersonal conflict such as domestic violence, road rage, assault, murder, and rape; and inter-group conflict including riots, ethnic violence, land invasions, gang violence, civil war and other forms of political instability, such as coups.
> (Burke, Hsiang, & Miguel, 2015: 1)

Based on the preceding definitions, this book considers climate change conflict as micro-level social and political conflict which is manifested as civil unrest, ethnic conflict, riots, killings, burning, protests, attacks, and human rights violations which are related to the effects of climate change. The manifestation of social conflict is varied. As Oberschall (1978: 291) argues, "social conflict encompasses a broad range of social phenomena: class, racial, religious, and communal conflicts; riots, rebellions, revolutions; strikes and civil disorders; marches, demonstrations, protest gatherings, and the like." Social conflict occurs between and among groups in a given society that compete over resources, social positions and values. Social conflict can be defined as "a struggle over values or claims to status, power, and scarce resources, in which the aims of the conflict groups are not only to gain the desired values, but also to neutralize, injure, or eliminate rivals" (Coser, 1967: 232).

Although the research endeavour to explore the connection between climate change and conflict is relatively recent, it has already produced many research reports and publications in well-respected journals (Figure 1.2). In 2007 and 2014 the journal *Political Geography* published special issues with several empirical research-based articles on the implications of climate change in conflict formation.[2] The *Journal of Peace Research* published a special issue on climate change and conflict in 2012 to address the security and conflict implications of climate change.[3] Most of these works have focused on the impacts of floods, droughts, and rainfall shortages on the social, economic, and political conditions of the worst affected countries (Figure 1.4). More recent works explore the relationship between climate change and conflict on a global scale, as well as local level case studies, to advance the theoretical understanding of climate change and conflict. Some researchers also point to climate change–induced migration as one of the sources of insecurity and conflict within and between countries when migrants put pressure on existing resources and when the political system in the host location fails to accommodate migrant people (Barnett & Adger, 2007; Goldstone, 2002; Scheffran et al., 2012).

In order to search the publication, the author used the key words *climate change and conflict relationship*. In summary, global climate change and its impacts on humans and the environment have constituted a critical area of discussion and research in various fields. The social dimensions of climate change concern population displacement, migration, insecurity, and conflict issues in the national and international context. Despite numerous publications and reports, there is still a lack of consensus on the key issue of recognizing climate displaced people and how climate migration transforms into a

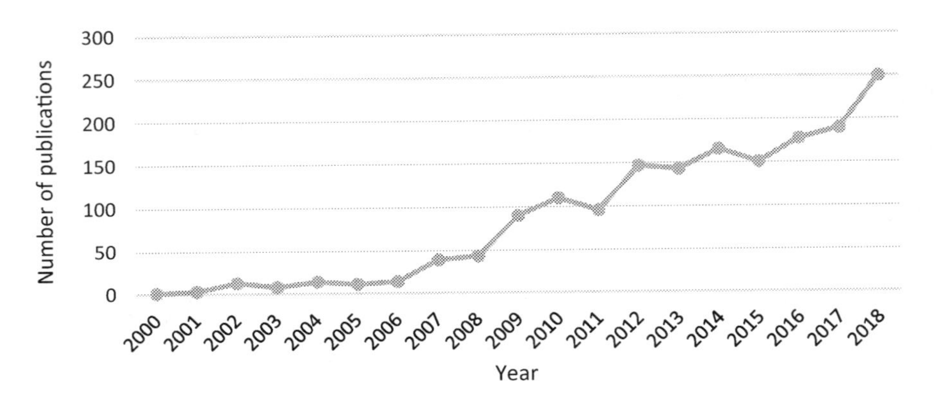

Figure 1.4 Trend in scholarly publications on climate change and conflict from 2000 to 2018

Source: Developed by the author based on published works in Google Scholar database, November 2018.

source of insecurity and conflict. Thus, further empirical research on climate change, migration, and conflict at the local level is deemed to be an important and timely endeavour in the academic and policy arena. Analyzing local case studies that explore the link between climate change, migration, and conflict is a step toward better understanding the connection between climate change and conflict.

Why Bangladesh as a case study

Bangladesh, as one of the most climate change–affected countries in the world, is experiencing climate change in the form of changing patterns of climatic events, namely "increasingly frequent and severe floods, cyclones, storm surges and drought forecasts" (Walsham, 2010). By climate change in Bangladesh, this thesis refers to both sudden and slow onset processes of environmental change, such as floods, cyclones, and riverbank erosion (sudden incidents), and coastal erosion, sea-level rise, saltwater intrusion, rising temperatures, changing rainfall patterns, and drought (slow processes). A combination of geographical location, poverty, weak infrastructure, lack of modern technology, and high population density has made the country particularly vulnerable to the risks associated with climatic events (Adger et al., 2003; Agrawala et al., 2003; Brouwer et al., 2007). Bangladesh is a small but densely populated country where 160 million people live on 167,598 square kilometres of land, which ranks Bangladesh as the 7th most densely populated country in the world (World Bank, 2020). The population density has forced people living in the low land and coastal areas to be exposed to the adverse effects of climatic events, such as floods, drought, sea-level rise, coastal erosion, and flood-driven river bank erosion.

Due to its location of the country at the top of the Bay of Bengal between 20–26° North and 88–92° East and its floodplain character, Bangladesh is affected every year by flooding, drought, and natural disasters (Agrawala et al., 2003). Bangladesh is also a riverine country, crisscrossed by more than 230 rivers. This riverine character brings both fortune and curse for the nation. During the monsoon season the country experiences widespread flooding, but in the dry season a significant part of the country, particularly the northeastern part, suffers from severe water shortage and drought (Mirza & Ahmad, 2005; Shamsuddoha & Chowdhury, 2007). More than 94 percent of inland water sources come from outside of the border from 54 international rivers, 51 flowing down from India and remaining three from Myanmar (Biswas, 2011; Elhance, 1999; Wirsing, 2007). India as a large neighbouring upper riparian country controls the water flow by constructing dams over the main international rivers (Mirza, 2005; Swain, 1996). This has given India considerable power to wreak environmental destruction in Bangladesh, as withholding water leads to desertification in the dry season (Elahi, 2001; Shahid & Behrawan, 2008), and releasing water in the rainy season can cause widespread flooding. Geographical location, socioeconomic condition, and overpopulation have made the country vulnerable to the risks and vulnerabilities associated with climate change events (Adger et al., 2003; Agrawala et al., 2003; Brouwer et al., 2007).

Climatic events accompanied by poor socioeconomic characteristics severely impact on human habitat, livelihood, and overall development of the country. The most far-reaching impact is the displacement of poor people from rural and coastal areas, which leads to internal and international migration. For instance, a survey in Hatia Island in Bangladesh found that "22% of households affected by tidal-surge floods, and 16% affected by riverbank erosion have moved to urban areas from low-land and rural areas" (Foresight, 2011: 13). Based on the severity of climatic events, Piguet (2008: 6) states that "Bangladesh is indeed an exceptional example of a country in which environment and climate change effects may be reaching a point where it derives significant human movement." At the Copenhagen Climate Summit in 2009, the Bangladeshi Minister of Environment and Forests warned that "twenty million people could be displaced [in Bangladesh] by the middle of the century" (Findlay & Geddes, 2011: 138).

In this context, it is surprising that there are no official government statistics on internal migration in Bangladesh. Scattered and sporadic research, based on migration models and projections, has published tentative internal migration figures such as 78 million people displaced by 2020 (CDMP II, 2014: 12); and 15 million people by 2050 due to climate change events (IUCN, 2015). Although there are debates about the numbers involved, there is little doubt that "Bangladesh, a low-lying country where agriculture is vulnerable to floods and salinization could end up accounting for one-third of South Asia's climate migrants" (Rigaud et al., 2018: 127).

There is also scant research on where displaced people move to and settle. It is argued that climate change–induced migrants generally move to nearby cities or major cities (e.g., Dhaka and Chittagong) for their livelihood and shelter (Afsar, 2003; Hassani-Mahmooei & Parris, 2012; Islam & Hasan, 2016; Kartiki, 2011). Bangladesh has attained enormous progress and development in the social and economic sectors. The bulk of the industrial growth is situated at the capital where the industrial production primarily takes place, and that growth is shared by the entire country (Mahmud, 2008). However, the fruits of this growth are often stunted by the frequent natural disasters such as coastal cyclones, floods, and draughts that plague the landscape of Bangladesh. The rural people, particularly those living in coastal areas and those that live in the peripheral regions of the country, are the most affected by these climatic events (Karim, 2018). For a chance at having better lives, these affected people move toward the industrial capital in search of opportunities which include Dhaka and Chattogram, and are often forced into living in unhospitable living conditions in the process (Islam, 2018; Mallick & Vogt, 2014).

The CHT is a hilly area located in the southeastern part in Bangladesh with a total land mass of 13,189 sq. km, constituting one-tenth of the total area of the country (see Figure 1.5) but 40 percent of the total forest of Bangladesh (Rasul, 2009). It is a distinct and unique region in Bangladesh where around 13[4] major ethnic groups live (Mohsin, 1997: 12). The culture, lifestyle, and patterns of livelihood of these ethnic minority groups are sharply different from the Bengali population which constitutes the majority in Bangladesh. Although the ethnic population has struggled for decades to be recognised as the indigenous people of the CHT, the government of Bangladesh (GoB) rejects the terms "*indigenous* or *Adibashi* and instead uses the term *small ethnic communities* in official discourse. In 2011, the committee entrusted with formulating the Constitutional Amendment proposed to include the term *small ethnic minorities* (khudra nrigoshti) in the Constitution of Bangladesh instead of *indigenous people* (Gerharz, 2014: 12). This is because acknowledging the CHT people as indigenous would complicate the status of the Bengali people, which the government claims have a history of 1000 years or more.

In order to avoid the politics of recognition, this thesis uses the term *tribal people* to refer to people who have been living in the CHT for at least 300 years. This term has also been employed by others researchers including Gurr and Lewis (Gurr, 1993: 50; Lewis, 2011: 28). The tribal people are also known as Jumma people as they depend on jhum cultivation for their livelihood. They are also viewed as the Pahari or hill people in the CHT (Adnan, 2004).

The CHT region experienced ethnopolitical conflict between the Shanti Bahini, and the Bangladesh army that lasted for more than two decades. The Shanti Bahini, an armed organization formed by Parbatya Chattagram Jana Samhati Samiti (PCJSS) in the CHT, launched a movement against the government of Bangladesh and the army with the aim of achieving the right to

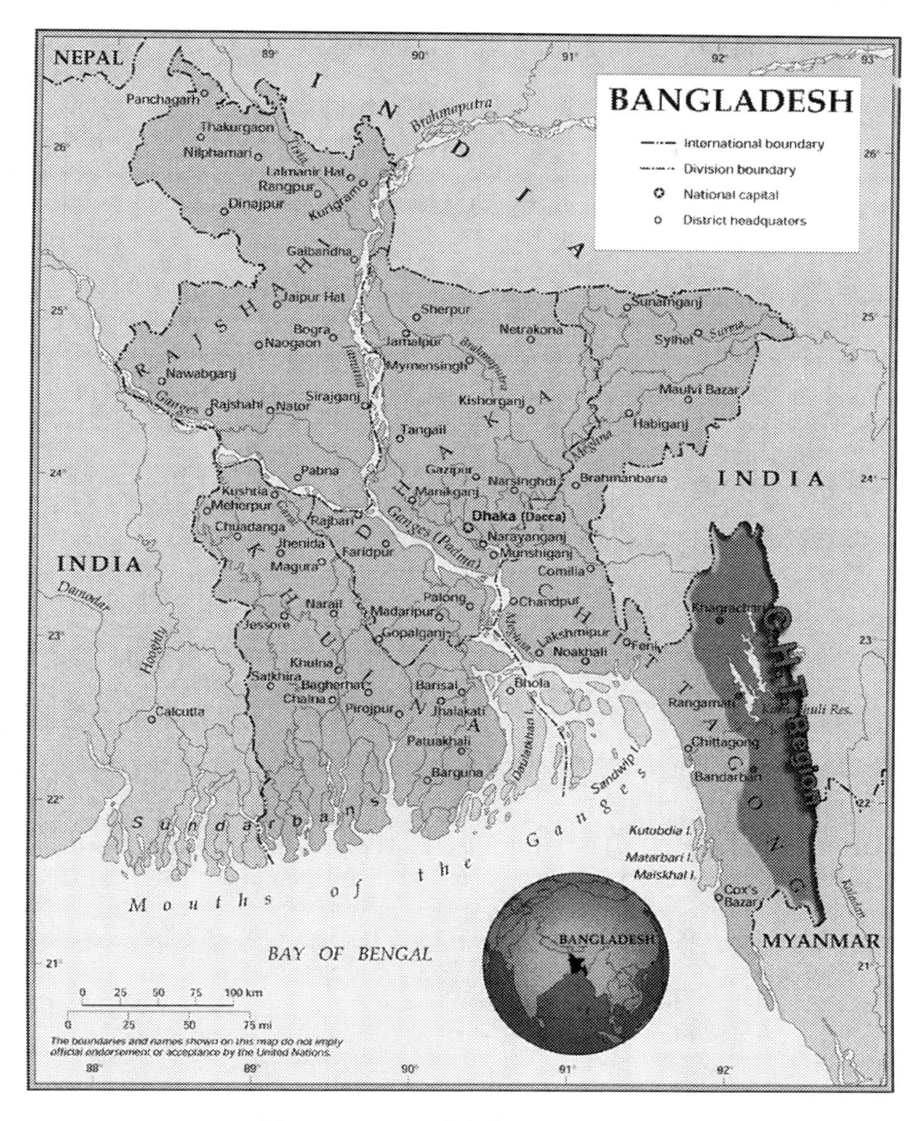

Figure 1.5 Location of the CHT in Bangladesh map

Source: Courtesy of Mohammed Kamal Hossain.

self-determination and recognition of identity (Chakma, 2010a). This ethno-political conflict was deemed to have ended with the signing of the Chittagong Hill Tracts Peace Accord in 1997 by the government of Bangladesh and the PCJSS, the indigenous political party led by Jyotirindra Bodhipriya Larma; however peace has remained elusive in the CHT (Jamil & Panday, 2008).

The CHT conflict is widely viewed in Bangladesh as caused by a combination of factors including denial of identity, military deployment, human rights violations, cultural annihilation, and economic dominance (Mohsin, 1997, 2000). However, the migration and settlement of the Bengali people to the CHT from the British colonial period to the present Bangladesh period has had significant impacts on the social, economic, and political conditions of the region. More importantly, the influx of Bengali population in the CHT under a settlement scheme (1977–1989) and subsequent informal migration flows has altered the demographic composition. Whereas the tribal people were the dominant population in the early 1970s, the numbers of Bengali and tribal people are now almost the same. This demographic transformation has created a complicated of social, economic and political relationships in the CHT. A new discourse of "Pahari-Bengali" (tribal and Bengali) has emerged in the CHT, which is influencing the socioeconomic and political situation in the post Peace Accord period (Nasreen, 2017: 139). This discourse refers to the fact that Bengali and tribal people have reached numeric equivalence and it is impossible now to ignore the Bengali community in social, economic, and political processes. Unfortunately, this demographic shift in the CHT has generated enmity, mistrust, and noncooperation between communities, termed micro-level social conflict by Choudhury, Islam, and Alam (2017). The communities are divided between the Bengali settlers and tribal people and engaged in conflict for resources and social position. Consequently, the socioeconomic and political situation in the CHT is one of conflict, despite various peacebuilding efforts which have been enacted in the post Peace Accord period such as the United Nations Development Programme Chittagong Hill Tracts Development Facility (UNDP-CHTDF), protection and promotion of human rights, capacity building of the local institutions, and community development (Chakma, 2017).

Researchers who examined environmental security and climate change issues in the CHT have suggested that Bengali settlement has been instrumental in causing resource scarcity and exploitation of tribal people and is, therefore, a factor in the CHT conflict (Hafiz & Islam, 1993; Lee, 1997, 2001; Reuveny, 2007, 2008; Reuveny & Moore, 2009; Suhrke, 1997). However, there has been no in-depth exploration of the complicated relationship between environmental destruction in lowland Bangladesh, population mobility, and conflict formation in the urban areas and CHT in Bangladesh. Van dar Hoorn's thesis on "Climate Change and Conflict in Bangladesh" (2010) is one of few studies on the topic, but it is based on only a few interviews with professional key informants and on secondary literature. Thus, the current study fills an important research gap by conducting empirical research to investigate the impacts of climatic events in the migration decisions of the people to the urban areas and CHT, and the impact of this migration on the conflict situation.

Outline of the book

This book comprises seven chapters. Chapter 1 explains the rationale of researching climate change, migration, and conflict issues. It paints an initial picture of why climate change is a significant issue for Bangladesh, why climate change induces migration at great scales in the country, and how this climate change–induced migration can become a source of harmful conflict for the country. By presenting the global picture of climate change, migration, and conflict, this chapter presents the Bangladesh case, as well as the CHT as a local level case study of climate change–induced migration and conflict. The exploration of the relationship between climate change–induced migration and conflict in Bangladesh can thus be seen as an important area, and this chapter demonstrates this.

Chapter 2 reviews the existing literature on climate change, migration, and conflict. This chapter critically explores the current debates of two separate but interlinking topics: the climate change and migration nexus; and climate change and conflict relationship. In the first part, this chapter analyzes the relationship between climate change and migration in order to explore a comprehensive overview of how climate change events affect people differently, in what circumstances the affected people take migration decisions, and where they usually settle as migrants. To analyze this connection, this section presents the intervening factors, also called mediating factors, which accomplish the migration process of climate affected people. Second, the chapter analyzes the relationship between climate change and conflict. This section explores when and how migration becomes a source of conflict and violence. The theoretical understanding of climate change, migration, and conflict relationship handled in this chapter will be a guiding outline to analyze the relationship between climate change induced migration and conflict in Bangladesh.

Chapter 3 provides a contextual study of Bangladesh by explaining relevant demographic information, climate change issues, and conflict history in Bangladesh. Based on secondary sources of information including population census data from the government of Bangladesh, this chapter provides a review of the existing literature on the demographic, socioeconomic, and climate change issues of the country. This chapter then provides a conflict history of Bangladesh, particularly in the CHT region – a region which was mainly inhabited by ethnic minority people. This chapter also explains how migration of Bengali people has changed the demographic character in urban areas and the CHT region of Bangladesh. It analyses the role of the government, social networks, and family connections in helping to settle the Bengali people in the CHT. A key contribution of this chapter is that existing research has ignored the implications of climatic events on migration flows to the CHT. This has led to the CHT conflict and climate–induced migration being studied in isolation.

Chapter 4 builds upon the findings of field work and a specific research question, which is how far the environment and climatic changes contributed to Bengali migration to the CHT and urban slums in Bangladesh. Having identified the major factors (climate change impacts and poverty) of the migration and settlement of the people, this chapter analyses the existing empirical studies and key informant interviews data to establish the role of climatic events in the migration decisions of people to the urban areas and the CHT in Bangladesh. In-depth analysis of interviews with climate migrants reveals how they have been affected by the floods, cyclones, and other environmental events in their place of origin, and why and under what conditions they decided to migrate to the urban areas and the CHT. This chapter contains key segments denoting the important contributions of this work in the literature of climate change and migration relationship in Bangladesh. The empirical findings from a local level case study have suggested how climatic events contributed to the migration decision, and also provide the complex relationship between climatic events and migration decision.

Chapter 5 analyses the interplay of climate change and migration relationship in Bangladesh with a focus on the mediating factors. The mediating factors are the contributing factors that enable the climate change–induced migration and settlement of impoverished people in the urban slums as well as in the CHT in Bangladesh. Population movement and migration from one place to another depends of contextual and mediating factors. This chapter finds that migration of the climate affected people to urban slums in Dhaka has occurred for the family connections, networks, and social connections. Based on historical information and interviews with key informants, this chapter discusses the mediating factors in the migration, which began in the British colonial era but has increased significantly since the creation of Bangladesh in 1971. Climate events, poverty, and political factors played a role in the migration to the CHT, which was institutionalized with the support of the government, security forces, and social networks.

Chapter 6 handles another important question of how migration of people with different backgrounds complicates the social, economic, and political situation in a host place. Climate change–induced migration of impoverished people to the urban slums puts pressure on the existing resources and services, which eventually triggers unrest, civil strife and violence in the city. On the other hand, Bengali migration to the CHT in different times has occurred for various reasons including the effects of climatic events. The main focus of this chapter is to explore the respondents' perceptions of the conflict situation in the CHT and how the Bengali migration and settlement have impacted the conflict situation. This chapter finds capturing resources, land grabbing, and competition for social and political positions between and among groups in the CHT region. This demographic shift due to the Bengali migration and polarization between the groups – ethnic and Bengali – has been a key factor to originate new conflicts and escalate the existing conflict. This chapter provides the important findings of this work based on the field data.

Chapter 7, the concluding chapter, provides a summary of the relationship of climate change, migration and conflict at the local level case study in Bangladesh. Summarizing the contextual study and two major findings chapters, the concluding chapter argues that the climate change, migration, and conflict relationship is not a liner; rather it is a complex relationship. No single effect can cause both the migration as well as conflict. Migration of a group people from one place to another depends on multiple factors, but climatic events play a pivotal role by complicating the existing environment in the highly affected locations. At the same time, the conflict formation due to migration is not always true. Different factors including the institutional and economic conditions are responsible for generating conflict in the places where migration takes place. The concluding chapter highlights this complex relationship of climate change, migration, and conflict based on the local level case study of Bangladesh.

Notes

1 This book has used *ethnic indigenous people*. There is a debate and discussion about the using the term *indigenous people* in Bangladesh. The tribal people living in the CHT claim themselves as the indigenous people. But the government of Bangladesh officially consider them as the small ethnic minority people. In order to avoid the confusion, this book has used the term *ethnic indigenous people*.
2 *Journal of Political Geography*, *43*, pp. 1–90 (November 2014), https://www.sciencedirect.com/journal/political-geography/vol/43/suppl/C; and Volume *26*(6), pp. 627–736 (August 2007), https://www.sciencedirect.com/journal/political-geography/vol/26/issue/6
3 *Journal of Peace Research* (January 2012; 49 (1). https://climateandsecurity.org/2012/02/02/climate-change-and-conflict-journal-of-peace-research-special-issue/
4 There are different views about the number of the tribal communities in the CHT. Some authors consider the number is 11, and some state 12 (Adnan, 2004; Mohsin, 1997).

References

Adamo, S. B. (2009). *Environmentally induced population displacements*. IHDP update, 1, 13–21. Bonn, Germany: International Human Dimensions Programme on Global Environmental Change. http://srdis.ciesin.columbia.edu/documents/environinduced-s.adamo-IHDPupdate-2009.pdf.

Adger, W. N. (2010). Climate change, human well-being and insecurity. *New Political Economy*, *15*(2), 275–292. https://doi.org/10.1080/13563460903290912.

Adger, W. N., Huq, S., Brown, K., Conway, D., & Hulme, M. (2003). Adaptation to climate change in the developing world. *Progress in Development Studies*, *3*(3), 179–195. https://doi.org/10.1191/1464993403ps060oa.

Adger, W. N., J. M. Pulhin, J. Barnett, G. D. Dabelko, G. K. Hovelsrud, M. Levy, Ú. Oswald Spring, & C. H. Vogel (2014). Human security. In:C. B. Field, V. R. Barros, D. J. Dokken, K. J. Mach, M. D. Mastrandrea, T. E. Bilir, M. Chatterjee, K. L. Ebi, Y. O. Estrada, R. C. Genova, B. Girma, E. S. Kissel, A. N. Levy, S. MacCracken, P. R. Mastrandrea, & L. L. White (Eds.), *Climate change 2014: Impacts, adaptation, and vulnerability* (pp. 755–791). Part A: Global and Sectoral Aspects. Contribution of Working Group II to the Fifth Assessment Report of the Intergovernmental Panel on Climate Change. Cambridge and New York: Cambridge University Press.

Adnan, S. (2004). *Migration land alienation and ethnic conflict: Causes of poverty in the Chittagong Hill Tracts of Bangladesh.* Dhaka: Research & Advisory Services.

Afifi, T., & Warner, K. (2008). *The impact of environmental degradation on migration flows across countries* (UNU-EHS Working Paper 5). Bonn: United Nations University Institute for Environment and Human Security. http://collections.unu.edu/view/UNU:1894

Afsar, R. (2003). *Internal migration and the development nexus: The case of Bangladesh.* Paper presented at the *Regional Conference on Migration, Development and Pro-Poor Policy Choices in Asia*, >22–24 June 2003 in Dhaka, Bangladesh. https://www.researchgate.net/publication/228916027_Internal_Migration_and_the_Devopment_Nexus_The_Case_of_Bangladesh

Agrawala, S., Ota, T., Ahmed, A. U., Smith, J., & Van Aalst, M. (2003). *Development and climate change in Bangladesh: Focus on coastal flooding and the Sundarbans* (pp. 1–49). Paris: Organisation for Economic Co-operation and Development (OECD). http://www.oecd.org/dataoecd/46/55/21055658.pdf

Ahsan, R., Kellett, J., & Karuppannan, S. (2016). Climate migration and urban changes in Bangladesh. In R. Shaw, A. Rahman, A. Surjan, G. Parvin (Eds.), *Urban disasters and resilience in Asia* (pp. 293–316). Amsterdam: Butterworth-Heinemann.

Baldwin, A. (2017). Climate change, migration, and the crisis of humanism. *Wiley Interdisciplinary Reviews: Climate Change*, *8*(3), 1–7. https://doi.org/10.1002/wcc.460.

Barnett, J. (2003). Security and climate change. *Global Environmental Change*, *13*(1), 7–17. https://doi.org/10.1016/S0959-3780(02)00080-8.

Barnett, J., & Adger, W. N. (2007). Climate change, human security and violent conflict. *Political Geography*, *26*(6), 639–655.

Bellard, C., Bertelsmeier, C., Leadley, P., Thuiller, W., & Courchamp, F. (2012). Impacts of climate change on the future of biodiversity. *Ecology Letters*, *15*(4), 365–377. https://doi.org/10.1111/j.1461-0248.2011.01736.x.

Biermann, F., & Boas, I. (2010). Preparing for a warmer world: Towards a global governance system to protect climate refugees. *Global Environmental Politics*, *10*(1), 60–88.

Biermann, F., & Boas, I. (2012). Climate change and human migration: Towards a global governance system to protect climate refugees. In J. Scheffran, M. Brzoska, H. G. Brauch, P. M. Link, & J. Schilling (Eds.), *Climate change, human security and violent conflict* (pp. 291–300). Berlin, Heidelberg: Springer.

Biswas, A. K. (2011). Cooperation or conflict in transboundary water management: Case study of South Asia. *Hydrological Sciences Journal*, *56*(4), 662–670.

Brouwer, R., Akter, S., Brander, L., & Haque, E. (2007). Socioeconomic vulnerability and adaptation to environmental risk: A case study of climate change and flooding in Bangladesh. *Risk Analysis*, *27*(2), 313–326. https://doi.org/10.1111/j.1539-6924.2007.00884.x

Brown, O., Hammill, A., & McLeman, R. (2007). Climate change as the "new"security threat: Implications for Africa. *International Affairs*, *83*(6), 1141–1154. https://doi.org/10.1111/j.1468-2346.2007.00678.x

Brzoska, M., & Fröhlich, C. (2016). Climate change, migration and violent conflict: Vulnerabilities, pathways and adaptation strategies. *Migration and Development*, *5*(2), 190–210. https://doi.org/10.1080/21632324.2015.1022973

Burke, M., Hsiang, S. M., & Miguel, E. (2015). Climate and conflict. *Annual Review of Economics*, *7*(1), 577–617. https://doi.org/10.1146/annurev-economics-080614-115430

Carleton, T. A., & Hsiang, S. M. (2016). Social and economic impacts of climate. *Science*, *353*(6304), aad9837. https://doi.org/10.1126/science.aad9837

Chakma, A. (2017). The peacebuilding of the Chittagong Hill Tracts (CHT), Bangladesh: Donor-driven or demand-driven? *Asian Journal of Peacebuilding*, *5*(2), 223–242.

Chakma, B. (2010). Structural roots of violence in the Chittagong Hill Tracts. *Economic and Political Weekly*, *45*(12), 19–21.

Chakma, B. (2010a). The post-colonial state and minorities: Ethnocide in the Chittagong Hill Tracts, Bangladesh. *Commonwealth & Comparative Politics*, *48*(3), 281–300. http://www.tandfonline.com/action/showCitFormats?doi=10.1080/14662043.2010.489746

Choudhury, Z. A., Islam, R., & Alam, S. (2017). Micro-foundation of conflict in the CHT. In Z. A. Choudhury (Ed.), *Conflict mapping in the Chittagong Hill Tracts (CHT) in Bangladesh* (pp. 52–126). Dhaka: Adhaorso Publication.

Christian Aid. (2007). *Human tide: The real migration crisis. A Christian Aid report*. London: Christian Aid. https://www.christianaid.org.uk/sites/default/files/2017-08/human-tide-the-real-migration-crisis-may-2007.pdf

Comprehensive Disaster Management Programme (CDMP II), (2014). *Trend and impact analysis of internal displacement due to the impacts of disaster and climate change*. Ministry of Disaster Management and Relief, Dhaka. https://reliefweb.int/sites/reliefweb.int/files/resources/CDMP-Internal-Displacement-Bangladesh-Analysis.pdf

Connell, J. (2003). Losing ground? Tuvalu, the greenhouse effect and the garbage can. *Asia Pacific Viewpoint*, *44*(2), 89–107. https://doi.org/10.1111/1467-8373.00187

Coser, L. A. (1967). *Continuities in the study of social conflict*. New York: Free Press.

Davenport, C. (2018, October 8). Major climate report warns of crisis by 2040. *The Newdaily*. https://thenewdaily.com.au/news/world/2018/10/08/climate-report-crisis-2040/.

Docherty, B., & Giannini, T. (2009). Confronting a rising tide: A proposal for a convention on climate change refugees. *Harvard Environmental Law Review*, *33*(2), 349–405.

Elahi, K. M. (2001). Drought in Bangladesh: A study of northwest Bangladesh. In K. Nazimuddin (Ed.), *Disaster in Bangladesh: Selected readings*. Dhaka: Disaster Research Training and Management Centre (DRTMC). Graphtone, Printers & Packaging Limited.

Elhance, A. P. (1999). *Hydropolitics in the Third World: Conflict and cooperation in international river basins*. Washington, DC: United States Institute of Peace Press.

Evans, A. (2011). Resource scarcity, climate change and the risk of violent conflict. *World Development Report, 2011*, Background paper. http://globalclimategovernance.org/sites/default/files/publications/ebaines/Resource%20scarcity,%20climate%20change%20and%20conflict.pdf

Faist, T. (2000). *The volume and dynamics of international migration and transnational social spaces*. Oxford: Clarendon.

Field, C. B. (2014). *Climate change 2014–impacts, adaptation and vulnerability: Regional aspects*. Cambridge: Cambridge University Press.
Findlay, A., & Geddes, A. (2011). Critical views on the relationship between climate change and migration: Some insights from the experience of Bangladesh. In E. Piguet, A. Pécoud, & P. de Guchteneire (Eds.), *Migration and climate change* (pp. 138–159). Paris: UNESCO Publishing.
Foresight. (2011). *Migration and global environmental change – future challenges and opportunities*. Final Project Report, The Government Office for Science, London: BIS. https://www.gov.uk/government/publications/migration-and-global-environmental-change-future-challenges-and-opportunities.
Gemenne, F., Barnett, J., Adger, W. N., & Dabelko, G. D. (2014). Climate and security: Evidence, emerging risks, and a new agenda. *Climate Change*, *123*(1), 1–9, Springer. https://doi.org/10.1007/s10584-014-1074-7.
Gerharz, E. (2014). Indigenous activism in Bangladesh: Translocal spaces and shifting constellations of belonging. *Asian Ethnicity*, *15*(4), 552–570.
Gleditsch, N. P. (2012). Whither the weather? climate change and conflict. *Journal of Peace Research*, *49*(1), 3–9. https://doi.org/10.1177%2F0022343311431288
Gleditsch, N. P., & Nordås, R. (2014). Conflicting messages? the IPCC on conflict and human security. *Political Geography*, *43*, 82–90. https://doi.org/10.1016/j.polgeo.2014.08.007
Gleditsch, N. P., Nordås, R., & Salehyan, I. (2007). *Climate change and conflict: The migration link. Coping with Crisis Working Paper Series*. New York: International Peace Academy. https://www.ipinst.org/2007/05/climate-change-and-conflict-the-migration-link
Gleick, P. H. (2014). Water, drought, climate change, and conflict in Syria. *Weather, Climate, and Society*, *6*(3), 331–340. https://doi.org/10.1175/WCAS-D-13-00059.1
Goldstone, J. A. (2002). Population and security: How demographic change can lead to violent conflict. *Journal of International Affairs*, *56*(1), 3–21. http://hdl.handle.net/1783.1/75219
Gurr, T. R. (1993). Why minorities rebel: A global analysis of communal mobilization and conflict since 1945. *International Political Science Review*, *14*(2), 161–201.
Hafiz, M. A., & Islam, N. (1993). *Environmental degradation and intra/interstate conflicts in Bangladesh*. ETH Zurich: Center for Security Studies (CSS).
Hassani-Mahmooei, B., & Parris, B. W. (2012). Climate change and internal migration patterns in Bangladesh: An agent-based model. *Environment and Development Economics*, *17*(6), 763–780. https://doi.org/10.1017/S1355770X12000290
Hendrix, C. S. (2018). Searching for climate–conflict links. *Nature Climate Change*, *8*(3), 190–191. https://doi.org/10.1038/s41558-018-0083-3
Hendrix, C. S., & Salehyan, I. (2012). Climate change, rainfall, and social conflict in Africa. *Journal of Peace Research*, *49*(1), 35–50. https://doi.org/10.1177%2F0022343311426165
Hinkel, J., Lincke, D., Vafeidis, A. T., Perrette, M., Nicholls, R. J., Tol, R. S., Marzeion, B., Fettweis, X., Ionescu, C., & Levermann, A. (2014). Coastal flood damage and adaptation costs under 21st century sea-level rise. *Proceedings of the National Academy of Sciences*, *111*(9), 3292–3297. https://doi.org/10.1073/pnas.1222469111
Homer-Dixon, T.F., Boutwell, J.H., Rathjens, G.W. (2011). Environmental change and violent conflict. In G. Machlis, T. Hanson, Z. Špirić, & J. McKendry (Eds.),

Warfare ecology: A new synthesis for peace and security, (pp.18–25). NATO science for peace and security series. Series C, environmental security (1874–6519). Dordrecht, London: Springer. https://doi.org/10.1007/978-94-007-1214-0_3.

Hugo, G. (2013). *Migration and climate change*. Cheltenham, UK: Edward Elgar Publishing.

Hunter, L. M. (2005). Migration and environmental hazards. *Population and Environment*, *26*(4), 273–302. https://dx.doi.org/10.1007%2Fs11111-005-3343-x

Ide, T., & Scheffran, J. (2014). On climate, conflict and cumulation: Suggestions for integrative cumulation of knowledge in the research on climate change and violent conflict. *Global Change, Peace & Security*, *26*(3), 263–279. https://doi.org/10.1080/14781158.2014.924917

Intergovernmental Panel on Climate Change (IPCC). (2007). *Contribution of working group II to the fourth assessment report of the intergovernmental panel on climate change: Summary of the policymakers*. Geneva: WMO, IPCC Secretariat. https://www.ipcc.ch/pdf/assessment-report/ar4/wg2/ar4-wg2-spm.pdf

Intergovernmental Panel on Climate Change (IPCC). (2014a). *Climate change 2014: Synthesis report*. Geneva 2, Switzerland Intergovernmental Panel on Climate Change. http://ipcc.ch/pdf/assessmentreport/ar5/syr/AR5_SYR_FINAL_All_Topics.pdf

Intergovernmental Panel on Climate Change (IPCC). (2014b). *Climate change 2014–impacts, adaptation and vulnerability: Regional aspects*. Cambridge: Cambridge University Press.

Internal Displacement and Monitoring Centre (IDMC). (2015). *The Global overview 2015: People internally displaced by conflict and violence*. Geneva: Switzerland. http://www.internal-displacement.org/publications/global-overview-2015-people-internally-displaced-by-conflict-and-violence

Internal Displacement and Monitoring Centre (IDMC). (2017). *Global report on internal displacemnt in 2016*. Geneva, Switzerland. http://www.internal-displacement.org/global-report/grid2017/pdfs/2017-IDMC-mini-Global-Report.pdf

IPCC. (2021). Climate change 2021: The physical science basis. Contribution of working group I to the sixth assessment report of the intergovernmental panel on climate change. In V. Masson-Delmotte, P. Zhai, A. Pirani, S. L. Connors, C. Péan, S. Berger ... & B. Zhou (Eds.), Geneva: Cambridge University Press.

Islam, M. R. (2018). Climate change, natural disasters and socioeconomic livelihood vulnerabilities: Migration decision among the Char land people in Bangladesh. *Social Indicators Research*, *136*(2), 575–593.

Islam, M. R., & Hasan, M. (2016). Climate-induced human displacement: A case study of cyclone Aila in the south-west coastal region of Bangladesh. *Natural Hazards*, *81*(2), 1051–1071.

Jamil, I., & Panday, P. K. (2008). The elusive peace accord in the Chittagong Hill Tracts of Bangladesh and the plight of the indigenous people. *Commonwealth & ComparativePolitics*,*46*(4),464–489.https://doi.org/10.1080/14662040802461141

Karim, A. (2018). The household response to persistent natural disasters: Evidence from Bangladesh. *World Development*, *103*, 40–59.

Kartiki, K. (2011). Climate change and migration: A case study from rural Bangladesh. *Gender & Development*, *19*(1), 23–38. https://doi.org/10.1080/13552074.2011.554017

Knox, J., Hess, T., Daccache, A., & Wheeler, T. (2012). Climate change impacts on crop productivity in Africa and South Asia. *Environmental Research Letters*, *7*(3), 034032.

Laczko, F., & Aghazarm, C. (2009). *Migration, environment and climate change: Assessing the evidence*. Geneva: International Organization for Migration. http://publications.iom.int/system/files/pdf/migration_and_environment.pdf.

Lee, S.-W. (1997). Not a one-time event: Environmental change, ethnic rivalry, and violent conflict in the Third World. *The Journal of Environment & Development*, 6(4), 365–396.

Lee, S.-W. (2001). *Environment matters: Conflicts, refugees & international relations*. Seoul and Tokyo: World Human Development Institute Press.

Lewis, D. (2011). *Bangladesh: Politics, economy and civil society*. Cambridge: Cambridge University Press.

Luetz, J. (2018). Climate change and migration in Bangladesh: Empirically derived lessons and opportunities for policy makers and practitioners. In W. Leal Filho & J. Nalau (Eds.), *Limits to climate change adaptation* (pp. 59–105). Cham: Springer.

Mach, K. J., Kraan, C. M., Adger, W. N., Buhaug, H., Burke, M., Fearon, J. D., Field, C. B., Hendrix, C. S., Maystadt, J.-F., O'Loughlin, J., Roessler, P., Scheffran, J., Schultz, K. A., & von Uexkull, N. (2019). Climate as a risk factor for armed conflict. *Nature*, *571*(7764), 193–197. https://doi.org/10.1038/s41586-019-1300-6

Mahmud, W. (2008). Social development in Bangladesh: Pathways, surprises and challenges. *Indian Journal of Human Development*, 2(1), 79–92.

Mallick, B., & Vogt, J. (2014). Population displacement after cyclone and its consequences: Empirical evidence from coastal Bangladesh. *Natural Hazards*, *73*(2), 191–212. https://doi.org/10.1007/s11069-013-0803-y

Matthew, R. A., Barnett, J., McDonald, B., & O'Brien, K. L. (2010). *Global environmental change and human security*. Cambridge, MA: MIT Press.

McAdam, J. (2010). *Climate change and displacement: Multidisciplinary perspectives*. Oxford, Portland: Hart Publishing.

McAdam, J. (2012). *Climate change, forced migration, and International law*. Oxford: Oxford University Press.

McLeman, R., & Smit, B. (2006). Migration as an adaptation to climate change. *Climatic Change*, 76(1–2), 31–53. https://doi.org/10.1007/s10584-005-9000-7

McLeman, R. A. (2014). *Climate and human migration: Past experiences, future challenges*. New York: Cambridge University Press.

McMichael, A. J., & Lindgren, E. (2011). Climate change: Present and future risks to health, and necessary responses. *Journal of Internal Medicine*, *270*(5), 401–413.

Mirza, M. M. Q., & Ahmad, Q. K. (Eds.). (2005). Climate change and water resources in South Asia: An Introduction. In M. M. Mirza & Q. K. Ahmed (Eds.), *Climate change and water resources in South Asia*, (pp. 1–21). Leiden, Netherlands: A.A. Balkema Pub.

Mohsin, A. (1997). *The politics of nationalism: The case of the Chittagong Hill Tracts*. Dhaka, Bangladesh: University Press Limited (UPL).

Myers, N. (1993). Environmental refugees in a globally warmed world. *Bioscience*, *43*(11), 752–761.

Myers, N. (2002). Environmental refugees: A growing phenomenon of the 21st century. *Philosophical Transactions of the Royal Society of London. Series B: Biological Sciences*, *357*(1420), 609–613.

Nasreen, Z. (2017). *The indigeneity question: State violence, forced displacement and women's narratives in the Chittagong Hill Tracts of Bangladesh*, Durham PhD theses, Durham University. Durham E-Theses Online. http://etheses.dur.ac.uk/12063/

Nordås, R., & Gleditsch, N. P. (2007). Climate change and conflict. *Political Geography*, *26*(6), 627–638.

Nordås, R., & Gleditsch, N. P. (2015). Climate change and conflict. In S. Hartard, & W. Liebert, (Eds.), *Competition and conflicts on resource use* (pp. 21–38). Heidelburg: Springer Cham.

Oberschall, A. (1978). Theories of social conflict. *Annual Review of Sociology*, *4*(1), 291–315.

Pachauri, R. K., Allen, M. R., Barros, V. R., Broome, J., Cramer, W., Christ, R., Church, J. A., Clarke, L., Dahe Dahe, Q., Dasgupta, P., Dubash, N. K., Edenhofer, O., Elgizouli, I., Field, C. B., Forster, P., Friedlingstein, P., Fuglestvedt, J., Gomez-Echeverri, L., Hallegatte, S., ... van Ypersele, J.-P. (2014). *Climate change 2014: Synthesis report. Contribution of working groups I, II and III to the fifth assessment report of the intergovernmental panel on climate change*. Geneva, Switzerland, IPCC.ISBN: 978-92-9169 143-2

Piguet, E. (2008). *Climate change and forced migration*. New Issues in Refugee Research, Research Paper No. 153, Policy Development and Evaluation Service, United Nations High Commissioner for Refugees. Geneva: Switzerland. http://www.unhcr.org/en-au/research/working/47a316182/climate-change-forced-migration-etienne-piguet.html

Pittock, A. B. (2017). *Climate change: Turning up the heat*. Routledge.

Rahman, A., 1999. Climate change and violent conflicts. In M. Suliman (Ed.), *Ecology, politics and violent conflict* (pp. 181–210). London and New York: Zed Books.

Raleigh, C., Jordan, L., & Salehyan, I. (2008). *Assessing the impact of climate change on migration and conflict*. Paper presented at the *World Bank Group for the Social Dimensions of Climate Change workshop*. Washington, DC: The World Bank. https://environmentalmigration.iom.int/assessing-impact-climate-change-migration-and-conflict

Rasul, G. (2009). Ecosystem services and agricultural land-use practices: A case study of the Chittagong Hill Tracts of Bangladesh. *Sustainability: Science, Practice and Policy*, *5*(2):15–27.

Reuveny, R. (2007). Climate change-induced migration and violent conflict. *Political Geography*, *26*(6), 656–673. https://doi.org/10.1016/j.polgeo.2007.05.001

Reuveny, R. (2008). Ecomigration and violent conflict: Case studies and public policy implications. *Human Ecology*, *36*(1), 1–13. https://doi.org/10.1007/s10745-007-9142-5

Reuveny, R., & Moore, W. H. (2009). Does environmental degradation influence migration? emigration to developed countries in the Late 1980s and 1990s*. *Social Science Quarterly*, *90*(3), 461–479. https://doi.org/10.1111/j.1540-6237.2009.00569.x

Rigaud, K. K., de Sherbinin, A., Jones, B., Bergmann, J., Clement, V., Ober, K., ... Heuser, S. (2018). *Groundswell: Preparing for internal climate migration*. Washington, DC: The World Bank.

Romm, J. (2018). *Climate change: What everyone needs to know?* New York: Oxford University Press.

Roy, D. C. (2011). Vulnerability and population displacements due to climate-induced disasters in coastal Bangladesh. In M. Leighton, X. Shen, & K. Warner (Eds.), *Climate change and migration: Rethinking policies for adaptation and disaster risk reduction* (pp.22–31). Bonn: United Nations University Institute for Environment and Human Security (UNU-EHS), 15.

Salehyan, I. (2008). From climate change to conflict? No consensus yet. *Journal of Peace Research*, *45*(3), 315–326. https://doi.org/10.1177%2F0022343308088812

Salehyan, I. (2014). Climate change and conflict: Making sense of disparate findings. *Political Geography*, *43*, 1–5, https://doi.org/10.1016/j.polgeo.2014.10.004

Scheffran, J., & Battaglini, A. (2011). Climate and conflicts: The security risks of global warming. *Regional Environmental Change*, *11*(1), 27–39.

Scheffran, J., Brzoska, M., Brauch, H. G., Link, P. M., & Schilling, J. (Eds.). (2012). *Climate change, human security and violent conflict: Challenges for societal stability*. Hexagon series on human and environmental security and peace. Berlin, New York: Springer.

Schwartz, P., & Randall, D. (2003). *An abrupt climate change scenario and its implications for United States national security*. Washington, DC: U.S. Department. of Defense, Environmental Media Services. https://eesc.columbia.edu/courses/v1003/readings/Pentagon.pdf

Selby, J., & Hoffmann, C. (2014). Rethinking climate change, conflict and security. *Geopolitics*, *19*(4), 747–756.

Shahid, S., & Behrawan, H. (2008). Drought risk assessment in the western part of Bangladesh. *Natural Hazards*, *46*(3), 391–413.

Shamsuddoha, M., & Chowdhury, R. K. (2007). *Climate change impact and disaster vulnerabilities in the coastal areas of Bangladesh*. Dhaka: COAST Trust.

Stapleton, S., Nadin, R., Watson, C., & Kellett, J. (2017). *Climate change, migration and displacement: The need for a risk-informed and coherent approach*. London, England: Overseas Development Institute. https://www.odi.org/sites/odi.org.uk/files/resource-documents/11874.pdf

Stern, N. (2006). *Stern review: The economics of climate change*. London: HM Treasury. http://mudancasclimaticas.cptec.inpe.br/~rmclima/pdfs/destaques/sternreview_report_complete.pdf

Suhrke, A. (1997) Environmental degradation, migration, and the potential for violent conflict. In N. P. Gleditsch (Es.), *Conflict and the environment* (pp. 255–272). NATO ASI Series (Series 2: Environment), Vol 33. Dordrecht: Springer.

Swain, A. (1996). Displacing the conflict: Environmental destruction in Bangladesh and ethnic conflict in India. *Journal of Peace Research*, *33*(2), 189–204.

Theisen, O. M., Gleditsch, N. P., & Buhaug, H. (2013). Is climate change a driver of armed conflict? *Climatic Change*, *117*(3), 613–625. https://doi.org/10.1007/s10584-012-0649-4

Theisen, O. M., Holtermann, H., & Buhaug, H. (2012). Climate wars? Assessing the claim that drought breeds conflict. *International Security*, *36*(3), 79–106. https://doi.org/10.1162/ISEC_a_00065

Trombetta, M. J. (2014). Linking climate-induced migration and security within the EU: Insights from the securitization debate. *Critical Studies on Security*, *2*(2), 131–147.

UNFCCC. (2007). *Climate change: Impacts, vulnerabilities and adaptation in developing countries*. Bonn: UNFCCC Secretariat. https://unfccc.int/resource/docs/publications/impacts.pdf

UNHCR. 2002. Environmental migrants and refugess. *Refugees*. No.127. http://www.unhcr.org./pub/PUBL/3d3fecb24.pdf

United Nations Framework Convention on Climate Change (UNFCCC), (1992). Definition, Article 1. http://unfccc.int/resource/ccsites/zimbab/conven/text/art01.htm

Urry, J. (2015). Climate change and society. In M. Jonathan, & C. Cary (Eds.), *Why the social sciences matter* (pp.45–59). Basingstoke: Palgrave Macmillan.
Walsh, K. J., McBride, J. L., Klotzbach, P. J., Balachandran, S., Camargo, S. J., Holland, G., Knutson, T. R., Kossin, J. P., Lee, T.-C., Sobel, A., & Sugi, M. (2016). Tropical cyclones and climate change. Wiley interdisciplinary reviews. *Climate Change*, 7(1), 65–89. https://doi.org/10.1002/wcc.371
Walsham, M. (2010). *Assessing the evidence: Environment, climate change and migration in Bangladesh*. Geneva: International Organization for Migration (IOM), Regional Office for South Asia. https://www.iom.int/jahia/webdav/site/myjahiasite/shared/shared/mainsite/events/docs/Assessing_the_Evidence_Bangaldesh.pdf
Warner, K. (2010). Global environmental change and migration: Governance challenges. *Global Environmental Change*, *20*(3), 402–413. https://doi.org/10.1016/j.gloenvcha.2009.12.001
WBGU, German Advisory Council on Global Change. (2008). *World in transition. Climate change as security risk*. London and Sterling, VA: Earthscan. https://www.wbgu.de/fileadmin/user_upload/wbgu.de/templates/dateien/veroeffentlichungen/hauptgutachten/jg2007/wbgu_jg2007_engl.pdf
Werz, M., & Conley, L. (2012). Climate change, migration, and conflict-addressing complex crisis scenarios in the 21st century. *Heinrich Boll Stiftung and Center for American Progress*. http://www.americanprogress.org
Wirsing, R. G. (2007). Hydro-politics in South Asia: The domestic roots of interstate river Rivalry. *Asian Affairs: An American Review*, *34*(1), 3–22.
World Bank. (2018). *Meet the human faces of climate migration*. https://openknowledge.worldbank.org/handle/10986/29461
World Bank. (2020). *Poverty and shared prosperity 2020: Reversals of fortune*. Washington, DC: The World Bank.

2 The climate change, migration, and conflict relationship

A review

Introduction

Climate change, migration, and conflict are three separate concepts which are used to present the climate change, migration, and conflict nexus in the social sciences literature. As a result, the literature, theoretical perspectives, and analytical lenses used to explain the climate change, migration, and conflict relationship are varied. In the 1970s and 1980s, researchers working in environmental change identified the potential for catastrophic effects of the environment on human beings. Many researchers during this time published reports and articles on climate change and migration (e.g., Myers, 1993, 2002; Hugo, 1996; El-Hinnawi, 1985), and the literature has grown significantly ever since. In contrast, the relationship between conflict and climate change is a relatively new area of study which emerged in the 1990s and was developed more broadly in the early 21st century (Deligiannis, 2012). Researchers from various disciplines have considered the political and social impacts of climate change and developed our theoretical and empirical understanding of the connection between climate change and conflict (Barnett & Adger, 2007; Reuveny, 2007; Nordås & Gleditsch, 2015; Barnett, 2003; Raleigh & Urdal, 2007). The relationship between climate change and migration posits that migration due to climate change may be forced or voluntary; temporary or permanent; internal or international; and sometimes is an adaptation strategy (Drabo & Mbaye, 2011; Hugo, 2013; Hummel, Doevenspeck, & Samimi 2012; Kartiki, 2011; McLeman, 2014). Climate change–induced migration is also seen through positive and negative perspectives. Such migration sometimes brings positive contributions to the host society through contributing to the economy and social development (IOM, 2010). Another positive lens portrays migration as a development strategy that increases social resilience in the face of mitigating climate vulnerabilities and shocks (Scheffran, Marmer, & Sow 2012; Webber & Barnett, 2010). However, climate change–induced migration may complicate the security of the host place and sometimes transform into a source of conflict. Since the beginning of

DOI: 10.4324/9781003430629-2

climate change and conflict research, many researchers have explored the impacts of increasing climatic events on security and the mechanisms leading to the formation of conflict in many locations across the world (Barnett & Adger, 2007; Burke, Hsiang, & Miguel, 2015; Hendrix & Salehyan, 2012; Hsiang & Burke, 2014; Raleigh & Urdal, 2007; Reuveny, 2007; Schleussner et al., 2016).

Despite much development in the theoretical and empirical understanding of climate change and migration, and climate change and conflict, there is still limited understanding of the relationships between these phenomena. In the case of the relationship between climate change and migration, researchers consider the causal relationship as underdeveloped, case and region-specific, and fragmented (Hugo, 2013; Piguet, Pécoud, & De Guchteneire, 2011; Bardsley & Hugo, 2010). Thus, it is suggested there is a need to conduct more empirical research based on the local level case study in order to better understand this relationship (McLeman, 2014). The theorization of the climate change and conflict relationship is also subject to criticisms on various grounds including methodology selection, causality analysis, regional bias, and the multicausal issues involved in most cases (Nordås & Gleditsch, 2007; Salehyan, 2008; Gleditsch, 2012; Buhaug, Gleditsch, & Theisen, 2010). Therefore, the relationship between climate change, migration, and conflict is recognized as a complex and debated issue that requires more empirical studies to explore how climate change leads to migration and conflict in different locations across the world (Gleditsch, 2012; McLeman, 2014; Hendrix, 2018).

For this reason, the literature review in this chapter aims to provide a broad understanding by discussing the multiple views on climate change and migration, as well the climate change, migration, and conflict relationship. It is presented in two parts: (1) the climate change and migration nexus, and (2) the climate change–induced migration and conflict relationship, to develop a framework for answering the research questions. The first section assesses the literature on how climatic events act as a driver of migration. This section discusses theoretical arguments on when and how people migrate in response to climatic changes, such as floods and sea-level rise, and what intervening or mediating factors accomplish the migration process. It also deals with the debates about the relationship between climate change and migration, in particular, whether climate migration is forced, voluntary, the outcome of government failure or an adaptation strategy. The second section summarizes and evaluates the current state of knowledge on how and under what conditions climate migration leads to conflict. Although a great deal of research focuses on different perspectives of climate change and conflict in general, this section only reviews literature that focuses on the mechanisms and processes by which climate change–induced migration can or has led to conflict. The final section addresses the major debates and the gaps in the research, as well as possibilities for further research.

Climate change and migration relationship

The general argument about the relationship between climate change and migration is that climatic events, both sudden and slow onset processes, complicate ecosystems and the human living spaces, agriculture, and livelihood options of millions of people in many locations in the world. Under such conditions, climate affected people are forced to migrate, or migrate voluntarily, to other locations. Climate change and environmental disasters force people to migrate by destroying their homes and livelihoods (Marino & Lazrus, 2015; Warner et al., 2010). In this sense, climatic events play a deterministic role in displacing people from their place of origin. This type of migration is often referred to as forced migration (Brown, 2007) as the affected people have no other choice but to migrate in search of livelihood. Sometimes climate affected people move to another location voluntarily in search of better opportunities and income. The affected people calculate the cost and benefits of migration from their homes. The climate change and migration relationship also posits that climate change events do not play the determining role in displacing people, but a number of factors such as mobility of people, the policy response by the government, uneven development, and socioeconomic status of a country influence the migration process (Raleigh & Jordan, 2010; Hugo, 2013; McLeman, 2014). Thus, existing explanations and frameworks include push, pull, and other factors, as well as adaptation, to understand the relationship between climate change and migration in any given place or situation. In relation to multiple causes, climate change and migration research has also incorporated social class, inequality, and mediating factors. The current research project draws from this basis and builds on this body of research.

The most common framework for analyzing the climate change and migration relationship invokes climatic events as push factors that render the environment uninhabitable for local people, reduce the resource base, and undermine the livelihood of people and consequently force them to migrate (Marino & Lazrus, 2015; Piguet, Pécoud, & De Guchteneire, 2011; Warner et al., 2010). As push factors, climatic events act as contextual determinants of migration (Burrows & Kinney, 2016) that drive people from their homes in the hope of finding security and livelihood options elsewhere (Perch-Nielsen, Bättig, & Imboden, 2008; Foresight, 2011). This migration often occurs reluctantly and in stages. When sudden climatic events displace people, they first seek safety as near as possible to their homes, often in temporary shelters. If they are unable to make a quick return to their home, they move a shorter distance as temporary migrants, hoping to return to their place of origin as soon as possible (Bardsley & Hugo, 2010; Tacoli, 2009; Raleigh & Jordan, 2010).

After destruction by climatic events, the short term shelter that is provided by the government does not usually enable people to maintain their lives for a longer period, particularly without assistance, and they need to find ways to make a living while they wait for the opportunity to return to their homes. Some factors such as adequate compensation, relief, and rehabilitation facilities from the government enable some migrants to return to their homes (Perch-Nielsen et al., 2008; Zaman & Wiest, 1991). The return to place of origin depends on the availability of houses, infrastructure development, government relief, and social capital (McLeman, 2014). For example, McLeman (2014: 103) explains that in New Orleans, USA post Hurricane Katrina, many people struggled to return to their homes due to a lack of housing, infrastructure, and relief. In the case of this inability, some people decide to move permanently to another place. Permanent migration mostly happens when the damage done to their home is much higher than the capacity of the people to return to home, or in other words, when it exceeds a threshold, which may vary from person to person (McLeman, 2018; Bardsley & Hugo, 2010). For example, in Bangladesh, floods and river bank erosion destroy agricultural land, houses, freshwater sources, and other means of livelihood, leading people to migrate to cities or to other rural areas (Mallick & Vogt, 2012; Perch-Nielsen et al., 2008).

Nevertheless, climate change does not always operate as a push factor forcing people to migrate. The capacity of people, community support, and timely intervention from the state sometimes prevents the damage from climatic events from forcing people to relocate and helps them to remain in their homes. This means the social, economic, and political conditions of an affected country determine the migration flow, either increasing migration or reducing it (Piguet, 2008: 3). Climatic events also do not force people to migrate suddenly; instead the sudden and gradual changes complicate the living place and livelihood options and make the place unlivable. Walsham (2010) outlines the four processes that determine climate change induced migration:

> intensifying natural disasters – both sudden and slow-onset – leading to increased displacement and migration; (b) shrinking and paralysing human security issues, such as: livelihoods, public health, food security and water availability; (c) rising sea level that make coastal areas uninhabitable; and (d) competition over scarce natural resources potentially leading to growing tensions and even conflict and, in turn, displacement.
>
> (Walsham, 2010: 5)

Permanent migration due to climatic events depends on contextual factors. Economically developed countries with sound technology and preparedness may soften the effects, reduce the cost and help people to remain in their homes despite causalities and destruction (Hugo, 2013; Raleigh & Jordan, 2010; Raleigh & Urdal, 2007). Social capital, community relationships, and resources also help affected people to mitigate the shock from climatic events. For example, McLeman (2014) considered that the social capital of the

Vietnamese community in New Orleans helped them to return to their home after Hurricane Katrina. The members of the community assisted their fellow people to return and resettle (McLeman, 2014: 101). However, in the case of poor economic conditions and weak technological capacity in underdeveloped and climate affected countries, people are more exposed to climate vulnerabilities. This is because affected people may not receive timely assistance to shift to safe places and then return to rebuild their houses. In summary, an analysis of push factors reveals that people migrate when they lose everything and fail to receive adequate support from the government, local institutions, and social networks.

Conversely, pull factors demonstrate that migration decisions of climate affected people are to some extent voluntary, and affected people may migrate to improve their economic position and overcome the hardships posed by climate vulnerability (Black, Kniveton, & Schmidt-Verkerk, 2013b; Bardsley & Hugo, 2010). The choices for employment opportunities, improving income, and reducing vulnerabilities influence the decisions of some climate affected people to migrate (Black et al., 2011; Fafchamps & Shilpi, 2013). In the wake of climatic events, affected people try to face the risk and overcome it through the available resources and social capital. Migrating or staying in the place of origin may depend on the behaviour of the affected people (Black, Adger et al., 2011a). Sometimes, family members, or at least one member, decide to migrate for more income so that the family left behind can overcome uncertainties (Beine & Parsons, 2015; Black, Arnell et al., 2013a). Many families in low-lying areas of Bangladesh have sent at least one family member abroad to improve their economic conditions via returned remittances and to reduce vulnerability (Siddiqui, 2010; Siddiqui & Billah, 2014). Families also try to send family members to the local cities for work opportunities. This migration pattern follows the human agency and livelihood approach, which outlines that in the face of climatic events, affected people receive material assistance and emotional support from the family members and already settled people in order to settle in a new place. This is called social networks and connections that motivate affected families to encourage a family member to migrate for income diversification (Hugo, 2008; Hummel, Doevenspeck, & Samimi, 2012). Social connection and support help to cushion the emotional impacts of being uprooted and facing significant personal challenges. For example, for farmers or agricultural labourers vulnerable to climatic events, migrating often involves shifting out of agricultural work and finding other employment in urban areas.

Some researchers propose population mobility as a solution because through migration people can improve their economic conditions (Warner, 2010; Reuveny, 2007). This migration is also seen as a positive phenomenon because it offers people the opportunity to settle in a safe place and reestablish their lives without having to face continuing climatic events (Gemenne & Blocher, 2017; McLeman, 2014). It is part of a wider trend of rural–urban migration as it is urban centers and industrial areas where migrants can find work opportunities (Tacoli, McGranahan, & Satterthwaite, 2015; Black, Adger, et al., 2011a). Some researchers argue that migration is an adaptation strategy for climate change instead of considering it as an impact (Bettini, 2014; Black, Bennett, et al., 2011b; Klaiber, 2014). From this perspective, migration is seen in a positive light rather than as a burden (Scheffran, Marmer, & Sow, 2012; Tacoli, 2009).

These contending ideas about climate change induced migration as burden or adaptation are connected to the debates on climate change–displaced people as refugees or migrants. As stated in the introduction of this chapter, some researchers and human rights organizations are eager to define climate change–induced mobility as refugee flows and call for international law to protect those involved (Biermann & Boas, 2010; McAdam & Saul, 2010). Against this effort, the UNHCR argues that climate migrants should not be labelled as refugees because they continue to enjoy national protection while political refugees do not, and thus need international protection (UNHCR, 2002: 13). Furthermore, recent research suggests that many climate affected people are unwilling to be labelled as refugees and do not wish to be resettled in other countries (Luetz & Havea, 2018).

Migration as an adaptation strategy has also been subject of debate. Different contending arguments have been developed over time. One of the most important arguments in favour of migration as an adaptation is that it enables people to save their lives by resettling in a safe location (de Sherbinin et al., 2011). In this context, McLeman (2014: 63) argues that "migration is a possible strategy by which those facing adverse climatic conditions may act to reduce the potential for loss or harm." This also applies to international migration, which provides migrants and their origin country with the remittances and other resources sent back to their place of origin. Migrants have the potentials to help the host country by sharing knowledge and technology and by providing services (World Bank, 2017).

Despite the arguments in favour of migration as an adaptation strategy, there are a number of counter-arguments. In the wake of climatic events, the affected people try to stay in their homes instead of adopting migration as a first act (Tacoli, 2009). A study based on the information of six million deidentified mobile phone users in Bangladesh after the cyclone Mahasen has found the weak correlation between the outmigration and effects of the cyclone (Lu et al., 2016). The worst affected people who lost everything migrated near towns near their place of origin. Permanent migration after the cyclone was very low, as most of the affected people returned to their place of origin after few months. This suggests that migration from one place to another depends on other issues such as financial cost, emotional attachment to homes and family members, and uncertainty (McLeman, 2014). Climate change migration is also seen as an outcome of the failure of the political system, government, and social networks. Unregulated migration can weaken the community and nation as a whole by imposing tremendous pressure on the existing labour forces and slowing development processes in the place of origin (Barnett & Campbell, 2010; Connell, 2003). In the face of growing migration, the affected area or country may face shortages of skilled labour which could slow their economic activity. Although climate migration may relieve the pressure on climate affected people and locations, it can also create resource scarcity, competition, crime, insecurity, and conflict in the host society (Blocher, 2016; Barnett & Adger, 2007; Reuveny, 2007, 2008). The migration of new groups may bring negative social, economic, and political consequences for multicultural or ethnic minorities and subsequently lead to social and political instability.

The relationship between climate change and migration is widely discussed as either a push, pull, or adaptation strategy; however, each framework has limitations and thus is not able to present a holistic picture. Thus, the relationship between climate change and migration is a complex process with multicausal issues (Hugo, 2013; McLeman, 2014). No single factor (push or pull) can determine migration, but composite factors such as networking, human agency, and demographic, political, and social issues contribute to the migration decisions of climate affected people (Hugo, 2013; McLeman, 2014; Piguet et al., 2011). People living in economically poor countries with high population density and social and political problems more often seek better living in other places (Black et al., 2008: 11). In this context, climate change effects can be an additional factor that accelerates the tendency to migration, because people seek livelihood and resources as well as deserve to save their life from disasters (Black et al., 2008). Frequent and sudden climatic events thus influence people to move to a safer place. Other factors such as social relations, dependence, assurance, and assistance open up the channels for migration (Hugo, 2013).

Government support and migration friendly policies also encourage climate affected people to migrate. In this sense, migration is seen as a "social product" (Hummel, Doevenspeck, & Samimi, 2012) where social, political, and economic issues influence the migration decisions of people. Family connections, links to people who have already migrated, government, and institutions represent a network that migrants can use to accomplish their migration process. Assurances by the government and institutional help at times assist people to migrate, even if the place they want to go to is adverse and situated a long distance away (Hugo, 2013). These other factors are known as intervening or mediating factors of migration and include transportation, networks, institutions, communication, and resources that influence people to migrate (Black, Adger et al., 2011a; Lilleør & Van den Broeck, 2011; McLeman, 2014; Reuveny, 2008). The role of mediating factors may accelerate or hinder population movement.

Some researchers invoke the role of social class and inequality when analyzing the climate change and migration relationship (Liu, 2015; Paavola & Adger, 2006). Poor and socially marginalized people are more prone to climate change migration. Climatic events adversely affect the poor, landless, and ethnic minority people. They also act to break social networks and family relationships when some people move to the cities while other family members remain in the place of origin. The poor economic conditions and social breakups lead some people to migrate permanently (Raleigh & Jordan, 2010: 112). The failure of the state system also plays a role in accelerating the migration process (Castles, 2002; Oliver-Smith, 2009). In less developed countries, corrupt practices by the public and NGO officials also affect the poor people because many times relief and assistance are not distributed to the affected people (Mahmud & Prowse, 2012). Instead, politicians and officeholders keep the resources for themselves. Corruption in any form affects disaster risk reduction, adaptation, and mitigation and also generates crimes and violence in the affected society (Alexander, 2017).

In previous sections, different frameworks have explained the relationship, mechanisms, and processes of climate change and migration, but in each framework, the issue of intervening and mediating factors is either overlooked or missing. Indeed, migration, whether it is regular or climate change induced, depends on mediating or intervening factors. The forced, voluntary, or adaptation analysis of climate change–induced migration is determined by a number of intervening or mediating factors (McLeman, 2014; Hugo, 2013). Overpopulation, income inequality, poverty, discrimination, and scarcity of resources are the contextual factors which become complicated in the face of adverse effects of climate change. These complex situations generate the migration situation; however, the process of migration is accomplished by mediating factors. These factors may include government policies, networks, family connections, and migration history, which influence the migration decisions of climate affected people (Brown, 2007; Raleigh & Jordan, 2010; Reuveny, 2007; Hugo, 2013; McLeman, 2014). Social networks, information, and family connections play a crucial mediating role in facilitating migration processes (Reuveny, 2008). Other mediating factors such as historical connections, cultural factors, behavioural attributes, and sociopolitical conditions help the aspirant people to make a decision to migrate (McLeman, 2014; Hugo, 2013). Security and political circumstances also act as mediating factors because governments and other political institutions control and promote migration for their own interests and benefit (McLeman, 2014). During the process of migration, some norms in migration pathways, values, and expectations among people who migrate are developed by mediating factors. Thus, mediating factors describe a diverse set of interrelated social, behavioural, institutional, and political factors that contribute to migration decisions which emerge from a combination of elements such as culture, geography, politics, economics, history, environment, and demography (Van Hear, Bakewell, & Long, 2018: 933).

It is clear from the oreceding discussion that the core idea of the climate change and migration nexus is that increasing climatic events in association with resource scarcity and social effects, such as poverty and economic underdevelopment, generate migration. Both sudden and slow onset climatic events create resource scarcity by diminishing resource stock and degrading resources. The people affected in such conditions struggle to manage their livelihood. They require assistance, relief, and government support to mitigate climate risk and remain in their homes, and when assistance and support are inadequate they migrate. Climate change displacement interacts with social and political circumstances such as poverty, discrimination, social exclusion, political instability, and corruption that encourages people to migrate from their place of origin. Climate migration is more often internal than international because many people hope to return to their homes following disasters, or maintain a close connection with family members left behind. As a whole, it is a complex process where multiple issues work together to determine the migration decision. This is illustrated in Figure 2.1.

Figure 2.1 shows that push, pull, and contextual factors influence the migration decision of the climate affected people. The behavioural factors also influence the migration decision. However, the climate affected people may have two options, such as migrate to another place or stay at their place of origin despite facing single or multiple factors. The affected people only migrate when they lose everything to survive, and they expect to have opportunities in the destination place. This means that the decision of migration depends on multiple factors.

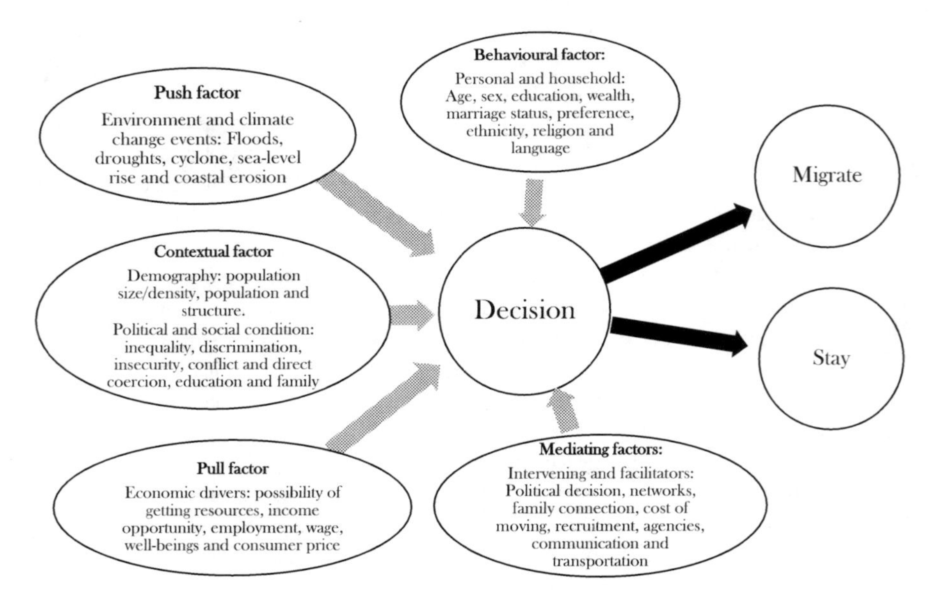

Figure 2.1 Causal relationships between climate change and migration with other drivers

Source: Based on the framework on climate change and migration relationship developed by Foresight (2011). This figure is adapted for the current study and may not be applicable in other case studies.

The nexus between climate change, migration, and conflict

How climate changes generate conflict is the second layer of analysis of climate change, migration, and conflict. The argument regarding the climate change and conflict relationship is that droughts, rainfall shortages, and temperature variations lead to water scarcity, failure of agriculture production, interruption to cattle rearing, and famine, which consequently lead to violence and conflict between and among groups (Gleick, 2014; Hendrix & Salehyan, 2012; Hsiang, Burke, & Miguel, 2013). The relationship between climate change and conflict is unpredictable and complex because there is no direct connection between the elements. Most studies have used quantitative studies to find the causality between global warming and increasing incidence of violence and conflict (Theisen, Gleditsch, & Buhaug, 2013; Ide & Scheffran, 2014). Moreover, the majority of the examples of climate change induced conflict are centered on Africa and the Middle East. Prolonged drought and rainfall shortages in those regions have reduced the economic activity and water sources. Water scarcity has consequently created situations of conflict in some locations between and among states in these regions (Bernauer & Siegfried, 2012; Feitelson, Tamimi, & Rosenthal, 2012). Research has been undertaken to explore how climate change and climate displaced people affect the indigenous people across the world. The indigenous and minority communities in various countries possess distinct characteristics and livelihood options (Ford et al., 2016). Climate change effects and the subsequent migration of different people intervene in the economy, social relations, and livelihood options (Adger et al., 2014).

The literature on the relationship between climate change and conflict emerged at the beginning of the 21st century; however, it has a connection with earlier literature from the 1990s demonstrating linkages between environmental resource scarcity and violence (Homer-Dixon, 1991, 1994; Homer-Dixon, Boutwell, & Rathjens, 1993). Environmental conflict refers to the political, social, economic, ethnic, religious, and territorial conflict that may cause resource scarcity, resource exploitation, and overuse (Baechler, 1998). Some researchers refer to national and human security issues in order to analyze the environmental conflict issue (Barnett, 2003; Barnett & Adger, 2007; Gleick, 1989). Environmental changes influence national and human security by paralyzing the vital resources on which state stability and human livelihood depend. Homer-Dixon (2007) in the *New York Times* (2007: A25) stated that "climate change events have potential to cause conflict and violence in the form of insurgency, genocide, guerrilla attacks, gang warfare, and global terrorism." The IPCC in its 5th assessment report also noted the potential for climate change to affect human security and generate conflict (Adger et al., 2014).

Nevertheless, the questions of how and what forms of conflict climate change generates, and under what conditions, still need to be explored. As mentioned in Chapter 1, conflict in the context of climate change tends to be small-scale, local level, and social, causing internal disturbances, civil strife, and ethnic rivalry. This definition of climate change and conflict correlates to the definitions of war, conflict, and environmental conflict, because parties, issues, and incompatibilities are also the key issues of analyzing climate change and conflict. Generally, conflict is defined by conflict researchers as a situation where two or more parties are engaged to achieve mutually incompatible goals (Mitchell, 1981: 17). The term *conflict* is often used synonymously with some a number of other terms, such as hostilities, tension, disharmony, struggle, or antagonism. Along with various meanings, the term *conflict* also takes various forms and scales, such as interpersonal, intercommunal, interstate, intrastate, group,and international conflict (Mitchell, 1981). Defining the relationship between climate change and conflict is an arduous task as most of the conflict issues are internal in nature. Conflict taking place within state borders is most likely to encompass multiple factors such as social, economic, government failure, ethnic division, and past record of conflict. This section consequently focuses on the theoretical understanding of climate change and conflict relations in general, and climate change–induced migration and conflict in particular.

The literature on the climate change and conflict relationship posits that climatic events have impacts on generating resource scarcity, which then become a catalyst for conflicts and violence (Homer-Dixon, 1999; Raleigh & Urdal, 2007; Barnett & Adger, 2007). Climate change in some locations indiscriminately affects people, reducing their resource base and the capacity of the environment on which humans depend. In such situations, some important social effects, such as resource scarcity and resource competition, develop, which eventually lead to conflict and violence. Resource scarcity only results in conflict when a zero-sum competition for the resources is organized between and among the groups. In such cases, some groups are totally deprived of their basic needs. Political authorities may fail to manage the scarce resources and accommodate all groups of people, and as a result, situations of conflict and violence occur (Homer-Dixon, 2010; Kahl, 2006). This explanation is part of the "neo-Malthusian" analysis which promulgates that population pressure on resources either from increasing population or migration leads societies closer to civil conflict and violence (Urdal, 2005; Hendrix & Glaser, 2007; Verhoeven, 2011). Countries heavily dependent on primary environmental resources are also more prone to conflict and violence if resources are degraded and depleted due to changes in precipitation and increasing temperatures, which are likely to be results of accelerated climate change (Homer-Dixon, 2010). McLeman (2014: 214) has presented the climate change–induced resource scarcity and conflict relationship shown in Figure 2.2.

According to Figure 2.2, people in any given society depend on resources which come from nature for their livelihood: climate change greatly affects the livelihood options and resources and consequently generates resource scarcity. Subsequently, resource scarcity and the interaction of individuals

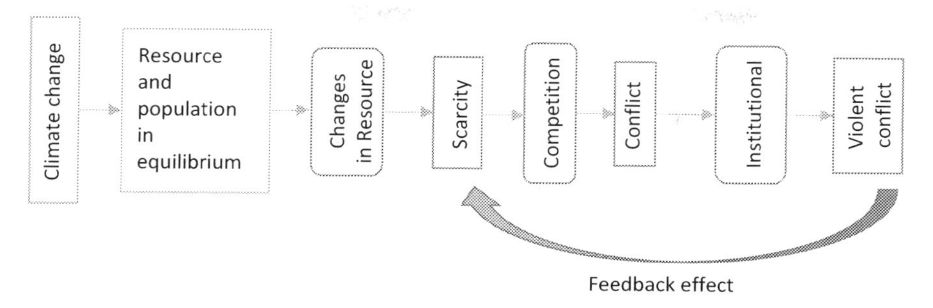

Figure 2.2 Relationship between resource scarcity and conflict
Source: (McLeman (2014: 214).

and groups sometimes results in conflict and violence. The explanation of climate change, scarcity, and conflict is based on qualitative and case studies which enabled the researcher to derive the underlying impacts of climate change in the social, economic, and political context. Although the nexus of climate change, resource scarcity, and conflict has a strong basis, it does not provide new explanations but has instead echoed the explanation of resource scarcity and violence developed in the 1990s (Buhaug et al., 2010). Moreover, there are shortcomings within the theory of this causal relationship of climate change, resource scarcity, and conflict. Some studies have found that resource scarcity has fewer impacts on generating conflict and violence in poor and undeveloped countries. Another study argues that scarcity of resources such as land and pasture has insignificant implications in generating armed conflicts (De Soysa & Neumayer, 2007). The local level arrangement of land distribution and conflict management mechanisms with effective institutions may help to allocate the existing resources among the contending groups and minimize the conflict (De Soysa, 2002). However, effective resource and conflict management in the face of resource scarcity is extremely difficult to achieve in poor and underdeveloped countries due to poverty, the behaviour of the political elites, and lack of knowledge among the people. Countries experiencing poverty, resource scarcity, and population pressure, as well as lack of ingenuity power (Homer-Dixon, 2010), are hardly able to accommodate all sections of the population. Under such conditions, competition and conflict inevitably emerge.

Some researchers state that climate change and disasters do not directly cause conflict, but rather, they function as a "threat multiplier" that triggers conflicts (Schleussner et al., 2016). Threat multipliers are the cumulative effects of climate change on social, economic, and political conditions (Scheffran, 2011). Poor countries are more prone to conflict in the face of increasing climate change due to their poor economy, dependence on subsistence agriculture, weak infrastructure, and poor record of human development (Brinkerhoff, 2011). This explanatory approach is known as a multicausal issue of conflict analysis. Climate and environmental stresses with other factors generate situations of conflict. Climate change as a threat multiplier affects the human security of people who already live in fragile places due to poverty, poor infrastructure, and high population density (Barnett, 2003; Barnett & Adger, 2007; Christiansen, 2016). Human security is generally defined as ensuring livelihood and life-supporting systems (Human Security Centre, 2005; Paris, 2001), and the sudden, or slow, onset of climate processes seriously undermines these systems. Population pressure, poverty, and poor governance in the climate affected regions or country mostly complicate the human security (Kahl, 2006; Brauch, 2014). The link between climate change, human security, and conflict is not a separate issue, but it is connected with analysis of how livelihood and scarcity of resources generate conflict. The failure of food and shelter and resource scarcity undermine the human security of the poor people who are forced to engage in competition for resources with contending groups. In such situations, resource capture by the powerful people pushes the poor people to be marginalized socially, economically, and environmentally, which may intensify human insecurity (Barnett & Adger, 2007; Adger et al., 2014).

Climate change–induced mass migration and the potential for conflict and violence are the strongest links in the analysis of the climate change and conflict nexus (Gleditsch, Nordas, & Salehyan, 2007). Climate migrants may generate conflict in the host place through complicating the existing social, economic, and political conditions (Null & Risi, 2016). In poor and underdeveloped societies, any form of migration puts pressure on the existing resources in the place of settlement. In such conditions, climate change–induced migrants compete with long-term residents in the host society for vital resources, such as land and forest, and this can become a source of conflict and violence (Raleigh & Urdal, 2007; Reuveny, 2007, 2008; Hendrix & Glaser, 2007). However, population mobility due to climate change is not the only factor to generate conflict; other factors such as resource capture and resource exploitation constitute an important ground for originating conflict (Homer-Dixon, 2010; Gleditsch, Nordas, & Salehyan, 2007). In some instances, government and political elites also compete for the resources and usurp them using their position of power (Reuveny, 2007: 659). This situation may generate conflict and violence when each group mobilizes against the other and opposes sharing resources (Reuveny, 2007, 2008). The role of government and institutions in such cases either defuses the conflict or instigates it by supporting one group over others. For example, in 2018 the BJP government in India strongly supported and provided assistance to the local Assamese people against the Bengali migrants, trying to deport a large number of migrants. The behaviour and policy of the government acted as a factor reigniting conflict between local people and migrants in the region (Bhaumik, 2018).

In some cases, the government and military may be engaged in exploiting resources and depriving local people who depend on them, which causes further frustration and conflict (Barnett, 2001). Underdeveloped countries are probable contexts for such conflict originating because governments are more likely to fail to accommodate climate change–induced migrants with livelihood and shelter. Furthermore, the government sometimes supports one group and pursues interests other than the greater public good because democratic processes in some developing countries are weak and governance systems corrupt. This situation is more prevalent where ethnic or minority people live and demand autonomy or the right to self-determination from the state. The government in such situations tend to suppress the minority or ethnic group, supporting other community members or groups through empowering and providing opportunities (Gurr, 1994, 2000). Stewart, Holdstock, and Jarquin (2002: 343) has termed this situation as "group inequality" where a particular group does not get access to political and economic opportunities due to the state mechanism. Newly emerging democratic countries with poor economic conditions and week institutions are most likely to become unable to accommodate the demands of the minority ethnic groups (Harff & Gurr, 2004).

Climate change–induced migration also contributes to conflict by altering the demographic composition of a country or a region, which can become a source of insecurity and conflict. The sudden migration of a large number of people weakens political legitimacy to rule the population if the government fails to provide basic services to people, and the host location may experience problems such as social breakdown, ecological collapse, and political and ethnic differences which trigger conflict (Adger et al., 2014; Verhoeven, 2011). Moreover, densely populated climate change affected countries are more prone to violence and conflict (Raleigh & Urdal, 2007: 675; Goldstone, 2002) as the increasing population puts excessive pressure on vital resources such as fresh water, land, food, and housing, and climate events can trigger civil unrest, violence, riots, and organized crime (Kahl, 1998, 2006). The increasing climate change effects destroy the natural resources as well as diminish the capacity of the resources to meet the demand of the people. The consequent situation greatly affects the social, economic, and political situation of the affected countries. Diamond (2005) argues that climate stress and overpopulation make the political system and government weak and vulnerable to collapse. Due to their inability to provide basic needs, people become desperate; seeing no hope for their future, they blame the government and become involved in illegal activities. People also see the government as responsible for their misfortune and behave irrationally: they fight, seize land from other people, and become desperate to migrate. Eventually, people become involved in serious conflict and violence (Diamond, 2005: 516). However, the connection of climate change, resource shrinking, population pressure, and outbreak of conflict may not happen in all climate affected regions. Good governance can manage the changing environment with the demand of the population. Moreover, this causal analysis has ignored the ability of the people to adapt to the changing environment and their inherent qualities to share the existing resources. Such adaptive capabilities most likely help people avoid conflict in the face of climate change (Hartmann, 2010: 237).

In the case of already existing conflict, climate change aggravates the conflict situation because it causes increased resource scarcity and population displacement. For example, many countries in the sub-Saharan region have a long history of ethnic and political conflict. Climatic events (prolonged drought, rainfall shortages, and increasing temperatures) have paralyzed the rural and regional economy and provided the trigger for civil strife. Political leaders have failed to allocate resources to the people. In these cases, climate change has played a role in increasing the existing conflict (Hendrix & Glaser, 2007; Hendrix & Salehyan, 2012; Raleigh, 2010). Hence, the existing conflict in the region has been aggravated due to the scarcity of resources, the inability of the government to provide basic needs, and competition for the resources among groups.

Climate change–induced migration may also cause insecurity and violence in an ethnically divided society. Some sources argue that the migration and conflict link is more visible if the area is inhabited by people with diverse cultural backgrounds (Barnett & Adger, 2007; Raleigh, 2010; Adger et al., 2014). Cultural practices, local knowledge, and lifestyles of the local ethnic people are affected by the influx of new migrants who bring their own cultural practices with them, and this generates tension between the established and new communities (Scheepers, Gijsberts, & Coenders, 2002; Weiner, 1992). When migration flows are large, they may also threaten the actual existence of ethnic minority people by interfering with their culture, resources, and livelihood practices (Dove, 2006; Swain, 1996; Goldstone, 2002). Reuveny (2007: 659) argues that the migration of new people may generate "ethnic tension" and "distrust" between and among the communities, and the consequent insecurity and fear of losing livelihood options can result in conflict and violence. For example, the conflict between Bengali migrants and tribal people in Assam in India is understood as environmental migration-induced conflict (Hazarika, 2000; Alam, 2003). Although the conflict in Assam is widely viewed as a political conflict, it has connection with environmentally induced migration of the Bengali people. The migration and settlement of the Bengali population to the CHT in Bangladesh has also escalated the conflicts by changing the ethnic composition of the region and increasing the rift, division, and polarization between the communities. Some researchers argue that environmentally induced migration of impoverished Bengali people mostly migrated to the CHT and intensified the tension and conflict in the region (Hafiz & Islam, 1993; Lee, 2001; Reuveny, 2007).

The connection between ethnic division and conflict at times depends on the behaviour of the state and treatment by the political authority of the minority people. Unequal treatment, hegemonic control, and deprivation of the ethnic minority community result in a sense of exploitation and marginalization (Clarke, 2001). In such cases, the ruling elites complicate the situation when they either deprive one group of people access to vital resources or facilitate another group's access to the same resources. However, developed countries such as Canada, Australia, and some European countries have been able to manage the migration issue effectively by careful resource allocation and distribution and enabling migrants to access required services (Castles, De Haas, & Miller, 2014). In contrast, poor people from the global South have less opportunity to migrate to developed countries in response to climate change. Moreover, the developed countries regulate the migration process to meet their demographic and economic needs. In such situations, actual conflict is absent between the host and migrant people. But there may still be conflict in the form of discrimination or the identification of new groups.

An important example of the climate change and conflict relationship is Darfur, Sudan. An 18-month study by the United Nations Environment Programme (UNEP) has explored how the drought in the Darfur region has reignited long-standing conflict. Rainfall shortages reduced crop yields and caused resource scarcity; as a result, the people from northern Darfur who were affected by the drought migrated to the southern and central areas in waves. After migration, they began to claim land for pastoral and agricultural use. This migration generated tension and conflict between the two groups (Borger, 2007; Kevane & Gray, 2008). This claim over land and livelihood gradually turned into a violent conflict. Although climatic events such as drought complicated agriculture and livelihood, the political failure, breakdown of national institutions, and role of national elites contributed to escalating the conflict (Verhoeven, 2011). For example, in the case of Darfur it has been argued that the involvement of the Sudanese government in supporting certain groups with arms and encouragement rather than climate change has been more instrumental in transforming competition over resources into a violent conflict. If the Sudanese government had instead made efforts to minimize or address the resource competition arising from migration into the region, this conflict may have been averted (Brown, Hammill, & McLeman, 2007). In this situation, different issues such as past conflict history, social divisions, and political failure, in addition to prolonged drought, contributed to developing conflict and escalating it to a new level. The example of climate change, migration, and conflict is similarly visible in other locations where climate change effects indiscriminately force people to migrate from their place of origin, and the destination place is not capable to accommodate the new migrant people. Different models and frameworks explain the relationship between climate change and conflict. But Figure 2.3 captures the relationship between climate change–induced migration and conflict more comprehensively.

Figure 2.3 shows that climate change events have implications for migration and resource scarcity. Migration itself has the potential to cause resource

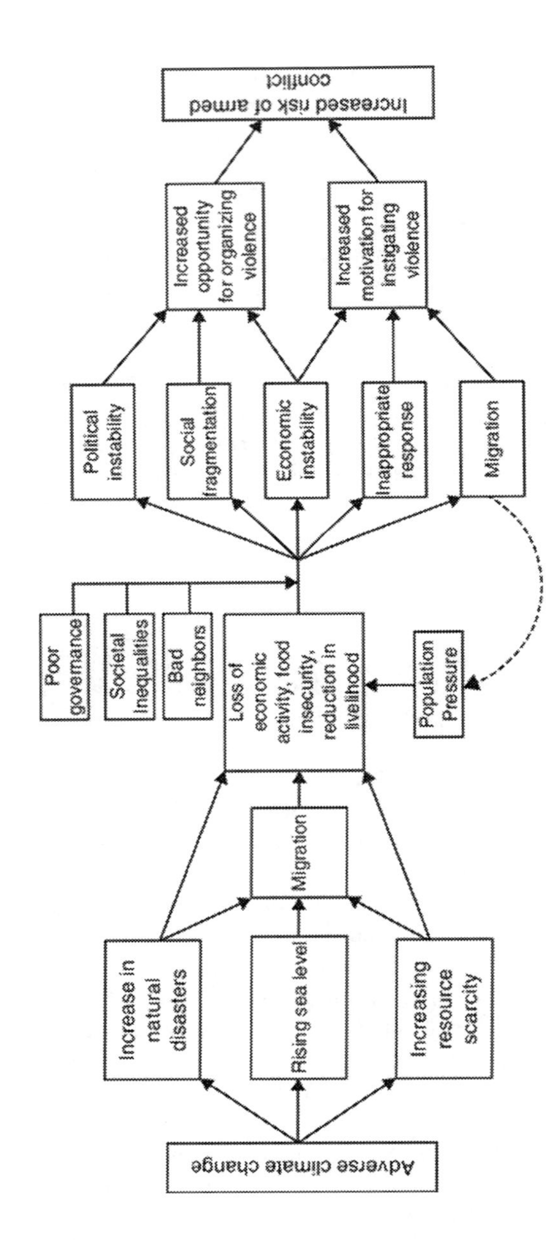

Figure 2.3 Possible pathways of climate change and conflict

Source: Buhaug et al. (2010: 82).

scarcity in the destination locations, and this means migration is viewed as the cause and effect of the conflict (Buhaug et al., 2010: 86). However, climate change–induced migration sometimes contributes to the economy of the host society. For example, migration to developed countries such as Australia does not cause resource scarcity; rather, the migration is carefully managed by the state and designed so that migrants contribute to the economy (Hugo, 2014). In the case of underdeveloped countries, it becomes extremely difficult to manage migration flows as the affected people need immediate shelter and food, and their movement may be unrestricted. The poor economic condition and climate change effects hinder the country from providing basic services to the citizens; thus they move to any suitable place. Climate change–induced migration in such a context most likely causes resource scarcity as the migrants put pressure on existing resources. Resource scarcity in association with sociopolitical and demographic factors may generate conflict situations in such societies where the social and political system fails to provide migrant people with services.

The discussion in this chapter connects three threads: climate change, migration, and conflict. At the first level of the relationship, climatic events generate daunting challenges such as resource scarcity, livelihood failure, and damage to agriculture and living places, all of which contribute to weakening the social and political system. The second layer of this relationship is migration which is caused by climatic events. A number of interconnected pathways such as destroying livelihood options, damage to houses and the displacement of poor people have characterized the climate change and migration relationship. Migration of poor people under these conditions is an inevitable issue as climate change forces people to migrate. The capacity for resilience and external support also influences whether or not affected people stay in their homes or migrate to another place for a better life. The last layer of the relationship is the conflict situation, which is the most unpredictable because conflict may not occur in all cases of climate change–induced migration. Conflict only occurs when climate change–induced migrations pose threats to the existing resources, challenge local and regional security issues, and change the demographic composition. In these situations, the relationship between climate change and conflict is indirect, where migration acts as a trigger to generate conflict. Migration also escalates existing conflicts through complicating the social, political, and economic conditions of the host place. This is sometimes referred to as a threat multiplier (Scheffran, 2011) because it exacerbates conflict in a society.

The relationship between climate change, migration, and conflict is clearly very complex and not fully understood (Gleditsch, 2012). Moreover, concrete evidence of climate change–induced conflict is lacking. To date, the studies that exist are regionally biased, as most of the empirical studies have been conducted in underdeveloped countries in Asia, Africa, and Latin America. Those regions have already experienced conflict and violence before being hit by extreme climatic events; thus drawing a general conclusion based on regional data and research output may not provide a comprehensive understanding of the climate change and conflict relationship (Nordås & Gleditsch, 2015). As these regions have a historical legacy of conflict and record of human rights violation, climate change effects in this context act not as the only factor but rather may act as triggering factors for conflict formation. Factors such as socioeconomic conditions, political structures, conflict history, and ethnic heterogeneity are responsible for originating conflicts in many poor countries. Resource scarcity sometimes encourages conflicting parties to join negotiations and cooperate in sharing resources instead of engaging in conflict (Dinar, 2009: 809). Good government can manage the scarcity of resources as well as manage the frustration of people due to resource scarcity.

Conclusion

In conclusion, it can be argued that the academic literature exploring the climate change, migration, and conflict relationship is vast, complex, and manifold. From this review, it is seen that the climate change and migration nexus has received a great deal of scholarly and policy attention which outlines that climate change effects, in connection with other forces such as pull influences and networks, sometimes force people to migrate. The core issue is the displacement from the place of origin to another location for livelihood and shelter. In fact, climate change–induced migration is an old phenomenon. Over different periods of time people have abandoned their homes and migrated to another place to overcome adverse environmental conditions (McLeman, 2014); however, since the middle of the 20th century, rapid climate changes and their associated events have displaced more people from their homes than in any other period. Thus, researchers and policy analysts have called for more empirical research to explore the complex and interconnected issues of climate change and migration. On the other hand, the linkages between climate change, migration, and conflict are a comparatively new research issue and still in a formative phase due to lack of established cases, evidence, theories, and frameworks that link accelerated climate change to conflict. This is because the nexus of climate change, migration, and conflict is explored in varied ways applying multiple methods and techniques. For example, qualitative approach and case studies have been applied to explore how resource scarcity may generate conflict and violence in different locations (Homer-Dixon, 2010). Some studies have applied regression analysis to find the direct link between climate change and conflict (Hauge & Ellingsen, 1998; Hendrix & Glaser, 2007); however, all these approaches and models are subject to criticism because they struggle to provide accurate causal linkages between climate change and conflict. Thus, some researchers turn to broader analysis of the climate change and conflict relationship incorporating social, political, and economic conditions with climate impacts. This multicausal analysis considers climatic events in connection with social, political, and economic factors which complicate the demographic and sociopolitical variables that eventually form conflict (Salehyan, 2014).

The literature review also reveals that researchers explore the relationship between climate change and migration, and climate change and conflict, separately. In both cases, some researchers are skeptical and critical about causal relationships. Therefore, it is suggested that in both cases there is a need to conduct more empirical research to substantiate the findings. Surprisingly, very little research and few studies have integrated the climate change, migration, and conflict relationship to explore the complex relationship and underlying mechanisms. This is one of the areas which needs investigation to see how climate change and migration, as well as climate change–induced migration and conflict, operate at a single local level. A local level case study based on empirical research should provide a more robust research result to substantiate the linkages between climate change, migration, and conflict (Hendrix, 2018).

The literature review also illustrates that different approaches and methods have been applied to see the relationships between climate change, migration, and conflict. Most of the quantitative studies have focused on the causality link between climate changes and occurrences of violence and conflict. These studies have found strong, moderate, or weak relationships between these elements (Hsiang, Burke, & Miguel, 2013; Hendrix & Salehyan, 2012); however, there is no comprehensive theory or understanding of this connection because evidence of climate change conflict is mostly focused on the African region. In this context detailed field work, particularly in other climate hotspots, would assist in understanding how climate change and migration operate as threat multipliers and how the climate change–induced migrants behave in the host place. The combining of methods, or mixed methods (Creswell & Clark, 2017), may assist in comprehending the complexity of climate change, migration, and conflict relationships. The conflict generating aspect of climate change and migration is also unexplored, especially if migration happens in a place where there is prior resource scarcity and the host society is already in conflict regarding their identity and self-determination. Thus, a local level study would be an important source for developing knowledge about climate change and conflicts as well as developing management approaches to mitigate the issues which emerge from migrant settlement.

In seeking to addressing this research gap of how climate change, migration, and conflict operate at a local level, the current book on Bangladesh is an endeavour to see how climate changes contributed to migration and conflict. This study is important as the CHT region is already conflict-prone, and Bangladesh is one of the most climate affected countries in the world. The next chapter provides the context in Bangladesh for exploring the impact of climate change in the migration decision of people, as well as how the migration process contributes to the conflict.

References

Adger, W. N., Pulhin, J. M., Barnett, J., Dabelko, G. D., Hovelsrud, G. K., Levy, M., Oswald Spring, Ú., & Vogel, C. H. (2014). Human security. In C. B. Field, V. R. Barros, D. J. Dokken, K. J. Mach, M. D. Mastrandrea, T. E. Bilir, M. Chatterjee, K. L. Ebi, Y. O. Estrada, R.C. Genova, B. Girma, E. S. Kissel, A. N. Levy, S. MacCracken, P. R. Mastrandrea, ... & L. L. White (Eds.), *Climate change 2014: Impacts, adaptation, and vulnerability* (pp. 755–791). Part A: Global and Sectoral Aspects. Contribution of Working Group II to the Fifth Assessment Report of the Intergovernmental Panel on Climate Change. Cambridge and New York: Cambridge University Press.

Alam, S. (2003). Environmentally induced migration from Bangladesh to India. *Strategic Analysis*, *27*(3), 422–438.

Alexander, D. (2017). Corruption and the governance of disaster risk. In S. L. Cutter (Ed.), *Oxford research encyclopedia of natural hazard science*. Oxford University Press, UK. http://oxfordre.com/naturalhazardscience/view/10.1093/acrefore/9780199389407.001.0001/acrefore-9780199389407-e-253?print=pdf

Baechler, G. (1998). Why environmental transformation causes violence: A synthesis. *Environmental Change and Security Project Report*, *4*(1), 24–44.

Bardsley, D. K., & Hugo, G. J. (2010). Migration and climate change: Examining thresholds of change to guide effective adaptation decision-making. Population and Environment, 32(2–3), 238–262. https://doi.org/10.1007/s11111-010-0126-9

Barnett, J. (2001). *The meaning of environmental security: Ecological politics and policy in the new security era*. London: Zed Books.

Barnett, J. (2003). Security and climate change. *Global Environmental Change*, *13*(1), 7–17. https://doi.org/10.1016/S0959-3780(02)00080-8

Barnett, J., & Adger, W. N. (2007). Climate change, human security and violent conflict. *Political Geography*, *26*(6), 639–655.

Barnett, J., & Campbell, J. (2010). *Climate change and small island states power, knowledge, and the South Pacific*. London: Earthscan.

Beine, M., & Parsons, C. (2015). Climatic factors as determinants of international migration. *The Scandinavian Journal of Economics*, *117*(2), 723–767.

Bernauer, T., & Siegfried, T. (2012). Climate change and international water conflict in Central Asia. *Journal of Peace Research*, *49*(1), 227–239.

Bettini, G. (2014). Climate migration as an adaption strategy: De-securitizing climate-induced migration or making the unruly governable? *Critical Studies on Security*, *2*(2), 180–195. https://doi.org/10.1080/21624887.2014.909225

Bhaumik, S. (2018, December 11). Deportation and detention: Three million face statelessness in Assam. *South China Morning Post*. https://www.scmp.com/week-asia/politics/article/2177296/indias-rohingya-three-million-face-statelessness-assams-crackdown

Biermann, F., & Boas, I. (2010). Preparing for a warmer world: Towards a global governance system to protect climate refugees. *Global Environmental Politics*, *10*(1), 60–88.

Black, R., Adger, W. N., Arnell, N. W., Dercon, S., Geddes, A., & Thomas, D. (2011a). The effect of environmental change on human migration. *Global Environmental Change*, *21*(1), S3–S11. https://doi.org/10.1016/j.gloenvcha.2011.10.001

Black, R., Arnell, N. W., Adger, W. N., Thomas, D., & Geddes, A. (2013a). Migration, immobility and displacement outcomes following extreme events. *Environmental Science & Policy*, *27*(1), S32–S43. https://doi.org/10.1016/j.envsci.2012.09.001

Black, R., Bennett, S. R., Thomas, S. M., & Beddington, J. R. (2011b). Climate change: Migration as adaptation. *Nature*, *4780*(7370), 447–449.

Black, R., Kniveton, D., & Schmidt-Verkerk, K. (2013b). Migration and climate change: Toward an integrated assessment of sensitivity. In T. Faist & J. Schade (Eds.), *Disentangling migration and climate change: Methodologies, political discourses and human rights* (pp. 29–53). Dordrecht: Springer Netherlands.

Black, R., Kniveton, D., Skeldon, R., Coppard, D., Murata, A., & Schmidt-Verkerk, K. (2008). *Demographics and climate change: Future trends and their policy implications for migration*. Working Paper, Development Research Centre on Migration, Globalization and Poverty, University of Sussex, Brighton. http://www.migrationdrc.org/publications/working_papers/WP-T27.pdf

Blocher, J. (2016). Climate change and environment related migration in the European Union policy: An organizational shift towards adaptation and development. In K. Rosenow-Williams & F. Gemenne (Eds.), *Organizational perspectives on environmental migration* (pp. 56–74). Abingdon, Oxon, New York: Routledge.

Borger, J., (2007, June 24). Darfur conflict heralds era of wars triggered by climate change. *The Guardian*. https://www.theguardian.com/environment/2007/jun/23/sudan.climatechange

Brauch, H. G. (2014). From climate change and security impacts to sustainability transition: Two policy debates and scientific discourses. In Ú. Oswald Spring, H. Brauch, & K. Tidball (Eds.), *Expanding peace ecology: Peace, security, sustainability, equity and gender*. SpringerBriefs in Environment, Security, Development and Peace, *12* (pp. 33–61). Heidelberg: Springer. https://doi.org/10.1007/978-3-319-00729-8_2

Brinkerhoff, D. W. (2011). State fragility and governance: Conflict mitigation and subnational perspectives. *Development Policy Review*, *29*(2), 131–153.

Brown, O. (2007). Climate change and forced migration: Observations, projections and implications. *UNDP Human Development Report, 2007/2008*, Occasional Paper 2007/17, 1, 3–34. https://www.iisd.org/library/climate-change-and-forced-migration-observations-projections-and-implications

Brown, O., Hammill, A., & McLeman, R. (2007). Climate change as the "new"security threat: Implications for Africa. *International Affairs*, *83*(6), 1141–1154. https://doi.org/10.1111/j.1468-2346.2007.00678.x

Buhaug, H., Gleditsch, N. P., Theisen, O. M., Mearns, R., & Norton, A. (2010a). Implications of climate change for armed conflict. In R. Mearns & A. Norton (Eds.), *Social dimensions of climate change: Equity and vulnerability in a warming world* (pp. 75–101). Washington, DC: The World Bank. http://siteresources.worldbank.org/INTRANETSOCIALDEVELOPMENT/Resources/SDCCWorkingPaper_Conflict.pdf

Buhaug, H., Gleditsch, N. P., Theisen, O. M., Mearns, R., & Norton, A. (2010b). Implications of climate change for armed conflict. In R. Mearns & A. Norton (Eds.), *Social dimensions of climate change: Equity and vulnerability in a warming world* (pp. 75–101). Washington, DC: The World Bank. http://siteresources.worldbank.org/INTRANETSOCIALDEVELOPMENT/Resources/SDCCWorkingPaper_Conflict.pdf

Burke, M., Hsiang, S. M., & Miguel, E. (2015). Climate and conflict. *Annual Review of Economics*, *7*(1), 577–617. https://doi.org/10.1146/annurev-economics-080614-115430

Burrows, K., & Kinney, P. L. (2016). Exploring the climate change, migration and conflict nexus. *International Journal of Environmental Research and Public Health*, *13*(4), 443.

Castles, S. (2002). Environmental change and forced migration: Making sense of the debate. *New Issues in Refugee Research*. Working Paper No-70, Geneva, Switzerland: United Nations High Commission for Refugee (UNHCR), I, 1–14.

Castles, S., De Haas, H., & Miller, M. J. (2014). *The age of migration: International population movements in the modern world* (4th ed.). New York: The Guilford Press.

Christiansen, S. M. (2016). *Climate conflicts: A case of international environmental and humanitarian law*. Cham: Springer International Publishing, Switzerland AG.

Clarke, G. (2001). From ethnocide to ethno-development? Ethnic minorities and indigenous peoples in Southeast Asia. *Third World Quarterly*, 22(3), 413–436.

Connell, J. (2003). Losing ground? Tuvalu, the greenhouse effect and the garbage can. *Asia Pacific Viewpoint*, *44*(2), 89–107. https://doi.org/10.1111/1467-8373.00187

Creswell, J. W. & Clark, V. P. (2017). *Designing and conducting mixed methods research* (3rd ed.). Thousand Oaks, CA: SAGE Publications.

de Sherbinin, A., Castro, M., Gemenne, F., Cernea, M., Adamo, S., Fearnside, P., Kriefer, G., Lahmani, S., Oliver-Smith, A., Pankhurst, A., Scudder, T., Singer, B., Tan, Y., Wannier, G., Boncour, P., Ehrhart, C., Hugo, G., Pandey, B., & Shi, G. (2011). Preparing for resettlement associated with climate change. *Science*, *334*(6055), 456–457.

De Soysa, I. (2002). Ecoviolence: Shrinking pie or honey pot? *Global Environmental Politics*, 2(4), 1–34.

De Soysa, I., & Neumayer, E. (2007). Resource wealth and the risk of civil war onset: Results from a new dataset of natural resource rents, 1970—1999. *Conflict Management and Peace Science*, *24*(3), 201–218.

Deligiannis, T. (2012). The evolution of environment-conflict research: Toward a livelihood framework. *Global Environmental Politics*, *12*(1), 78–100. https://doi.org/10.1162/GLEP_a_00098

Diamond, J. (2005). *Collapse: How societies choose to fail or succeed*. New York: Viking.

Dinar, S. (2009). Scarcity and cooperation along international rivers. *Global Environmental Politics*, *9*(1), 109–135. https://doi.org/10.1111/j.1468-2478.2011.00671.x

Dove, M. R. (2006). Indigenous people and environmental politics. *Annual Review of Anthropology*, *35*, 191–208. https://doi.org/10.1146/annurev.anthro.35.081705.123235

Drabo, A., & Mbaye, L. (2011). *Climate change, natural disasters and migration: An empirical analysis in developing countries*. Bonn: IZA Discussion Paper 5927.

El-Hinnawi, E. (1985). *Environmental refugees*. Nairobi, Kenya: United Nations Environment Programme (UNEP).

Fafchamps, M., & Shilpi, F. (2013). Determinants of the choice of migration destination. *Oxford Bulletin of Economics and Statistics*, *75*(3), 388–409. https://doi.org/10.1111/j.1468-0084.2012.00706.x

Feitelson, E., Tamimi, A., & Rosenthal, G. (2012). Climate change and security in the Israeli-Palestinian context. *Journal of Peace Research*, *49*(1), 241–257. https://doi.org/10.1177%2F0022343311427575

Ford, J. D., Cameron, L., Rubis, J., Maillet, M., Nakashima, D., Willox, A. C., & Pearce, T. (2016). Including indigenous knowledge and experience in IPCC assessment reports. *Nature Climate Change*, 6(4), 349–353.

Foresight. (2011). *Migration and global environmental change – future challenges and opportunities*. Final Project Report, The Government Office for Science, London: BIS. https://www.gov.uk/government/publications/migration-and-global-environmental-change-future-challenges-and-opportunities

Gemenne, F., & Blocher, J. (2017). How can migration serve adaptation to climate change? Challenges to fleshing out a policy ideal. *The Geographical Journal*, *183*(4), 336–347.

Gleditsch, N. P. (2012). Whither the weather? climate change and conflict. *Journal of Peace Research*, *49*(1), 3–9. https://doi.org/10.1177%2F0022343311431288

Gleditsch, N. P., Nordas, R., & Salehyan, I. (2007). Climate change and conflict: The migration link. Coping with Crisis Working Paper Series. *International Peace Academy, New York*. https://www.ipinst.org/2007/05/climate-change-and-conflict-the-migration-link

Gleick, P. H. (1989). Climate change and international politics: Problems facing developing countries. *Ambio*, *18*(6), 333–339.

Gleick, P. H. (2014). Water, drought, climate change, and conflict in Syria. *Weather, Climate, and Society*, *6*(3), 331–340. https://doi.org/10.1175/WCAS-D-13-00059.1

Goldstone, J. A. (2002). Population and security: How demographic change can lead to violent conflict. *Journal of International Affairs*, *56*(1), 3–21. http://hdl.handle.net/1783.1/75219

Gurr, T. R. (1994). Peoples against states: Ethnopolitical conflict and the changing world system: 1994 presidential address. *International Studies Quarterly*, *38*(3), 347–377.

Gurr, T. R. (2000). *Peoples versus states: Minorities at risk in the new century*. Washington, DC: United States Institute of Peace Press.

Hafiz, M. A., & Islam, N. (1993). *Environmental degradation and intra/interstate conflicts in Bangladesh*. ETH Zurich: Center for Security Studies (CSS).

Harff, B., & Gurr, T. R. (2004). *Ethnic conflict in world politics* (2nd ed.). New York: Westview Press.

Hartmann, B. (2010). Rethinking climate refugees and climate conflict: Rhetoric, reality and the politics of policy discourse. *Journal of International Development*, *22*(2), 233–246. https://doi.org/10.1002/jid.1676

Hauge, W., & Ellingsen, T. (1998). Beyond environmental scarcity: Causal pathways to conflict. *Journal of Peace Research*, *35*(3), 299–317.

Hazarika, S. (2000). *Rites of passage: Border crossings, imagined homelands, India's East and Bangladesh*. New York: Penguin Books.

Hendrix, C. S. (2018). Searching for climate–conflict links. *Nature Climate Change*, *8*(3), 190–191. https://doi.org/10.1038/s41558-018-0083-3

Hendrix, C. S., & Glaser, S. M. (2007). Trends and triggers: Climate, climate change and civil conflict in Sub-Saharan Africa. *Political Geography*, *26*(6), 695–715. https://doi.org/10.1016/j.polgeo.2007.06.006

Hendrix, C. S., & Salehyan, I. (2012). Climate change, rainfall, and social conflict in Africa. *Journal of Peace Research*, *49*(1), 35–50. https://doi.org/10.1177%2F0022343311426165

Homer-Dixon, T. F. (1991). On the threshold: Environmental changes as causes of acute conflict. *International Security*, *16*(2), 76–116.

Homer-Dixon, T. F. (1994). Environmental scarcities and violent conflict: Evidence from cases. *International Security*, *19*(1), 5–40.

Homer-Dixon, T. F. (1999). *Environment, scarcity, and violence*. Princeton: Princeton University Press.

Homer-Dixon, T. F. (2007, April 24), Terror in the weather forecast. *The New York Times*. https://www.nytimes.com/2007/04/24/opinion/24homer-dixon.html

Homer-Dixon, T. F. (2010). *Environment, scarcity, and violence*. New Jersey: Princeton University Press.

Homer-Dixon, T. F., Boutwell, J. H., & Rathjens, G. W. (1993). Environmental change and violent conflict. *Scientific American*, *268*(2), 38–45.

Hsiang, S. M., & Burke, M. (2014). Climate, conflict, and social stability: What does the evidence say? *Climatic Change*, *123*(1), 39–55.

Hsiang, S. M., Burke, M., & Miguel, E. (2013). Quantifying the influence of climate on human conflict. *Science*, *341*(6151), 1235367. https://www.semanticscholar.org/paper/Environmental-Change-%2C-Migration-and-Conflict-%3A-andReuveny/7aed090a3441dc28cb2c369b2de420acf3ea5ce1

Hugo, G. (1996). Environmental concerns and international migration. *International Migration Review*, XXX (1), 105–131.

Hugo, G. (2008). *Migration, development and environment*. IOM Migration Research Series, No-35. Geneva: International Organization for Migration. https://publications.iom.int/system/files/pdf/mrs_35_1.pdf

Hugo, G. (2013). *Migration and climate change*. Cheltenham, UK: Edward Elgar Publishing.

Hugo, G. (2014). Change and continuity in Australian international migration policy. *International Migration Review*, *48*(3), 868–890.

Human Security Centre. (2005). *Human security report 2005: War and peace in the 21st Century*. New York: Oxford University Press.

Hummel, D., Doevenspeck, M., & Samimi, C. (2012). Climate change, environment and migration in the sahel: Selected issues with a focus on Senegal and Mali. In D. Hummel, M. Doevenspeck, & C. Samimi (Eds.), *Migration, climate and environment* (pp. 1–81). MICLE Working Paper No-1. Frankfurt/Main. http://www.micle.project.net/uploads/media/micle-wp1-2012-en.pdf

Ide, T., & Scheffran, J. (2014). On climate, conflict and cumulation: Suggestions for integrative cumulation of knowledge in the research on climate change and violent conflict. *Global Change, Peace & Security*, *26*(3), 263–279. https://doi.org/10.1080/14781158.2014.924917

International Organization for Migration (IOM). (2010). *Migration, climate change and environmental degradation: Definitional issues*. Geneva: IOM. http://www.iom.int/jahia/Jahia/activities/by-theme/migration-climate-change-environmental-degradation/definitional-issues

Kahl, C. H. (1998). Population growth, environmental degradation, and state-sponsored violence: The case of Kenya, 1991–93. *International Security*, *23*(2), 80–119.

Kahl, C. H. (2006). *States, scarcity, and civil strife in the developing world*. Princeton; Woodstock: Princeton University Press.

Kartiki, K. (2011). Climate change and migration: A case study from rural Bangladesh. *Gender & Development*, *19*(1), 23–38. https://doi.org/10.1080/13552074.2011.554017

Kevane, M., & Gray, L. (2008). Darfur: Rainfall and conflict. *Environmental Research Letters*, *3*(3), 034006.

Klaiber, H. A. (2014). Migration and household adaptation to climate: A review of empirical research. *Energy Economics*, *46*, 539–547. https://doi.org/10.1016/j.eneco.2014.04.001

Lee, S.-W. (2001). *Environment matters: Conflicts, refugees & international relations*. Seoul and Tokyo: World Human Development Institute Press.

Lilleør, H. B., & Van den Broeck, K. (2011). Economic drivers of migration and climate change in LDCs. *Global Environmental Change*, *21*(1), S70–S81. https://doi.org/10.1016/j.gloenvcha.2011.09.002

Liu, X. (2015). *Exploring the relationship between climatic variability, inequality and migration from a class perspective: Evidence from Minqin County, Western China*, Unpublished PhD desertation, University of Adelaide, Australia. https://digital.library.adelaide.edu.au/dspace/bitstream/2440/98152/2/02whole.pdf

Lu, X., Wrathall, D. J., Sundsøy, P. R., Nadiruzzaman, M., Wetter, E., Iqbal, A., Qureshi, T., Tatem, A., Canright, G., Engø-Monsen, K., & Bengtsson, L. (2016). Unveiling hidden migration and mobility patterns in climate stressed regions: A longitudinal study of six million anonymous mobile phone users in Bangladesh. *Global Environmental Change*, *38*, 1–7.

Luetz, J., & Havea, P. H. (2018). We're not refugees, we'll stay here until we die!—climate change adaptation and migration experiences gathered from the Tulun and Nissan Atolls of Bougainville, Papua New Guinea. In W. L. Filho (Ed.), *Climate change impacts and adaptation strategies for coastal communities* (pp. 3–29). Cham, Switzerland: Springer.

Mahmud, T., & Prowse, M. (2012). Corruption in cyclone preparedness and relief efforts in coastal Bangladesh: Lessons for climate adaptation? *Global Environmental Change*, *22*(4), 933–943.

Mallick, B., & Vogt, J. (2012). Cyclone, coastal society and migration: Empirical evidence from Bangladesh. *International Development Planning Review*, *34*(3), 217–240. https://doi.org/10.3828/idpr.2012.16

Marino, E., & Lazrus, H. (2015). Migration or forced displacement?: The complex choices of climate change and disaster migrants in Shishmaref, Alaska and Nanumea, Tuvalu. *Human Organization* *74*(4), 341–350. https://doi.org/10.17730/0018-7259-74.4.341

McAdam, J., & Saul, B. (2010). Displacement with dignity: International law and policy responses to climate change mitigation and security in Bangladesh. *German Yearbook of International Law*, *53*, 233.

McLeman, R. (2018). Thresholds in climate migration. *Population and Environment*, *39*(4), 319–338. https://doi.org/10.1007/s11111-017-0290-2

McLeman, R. A. (2014). *Climate and human migration: Past experiences, future challenges*. New York: Cambridge University Press.

Mitchell, C. R. (1981). *The structure of international conflict*. London: Macmillan.

Myers, N. (1993). Environmental refugees in a globally warmed world. *Bioscience*, *43*(11), 752–761.

Myers, N. (2002). Environmental refugees: A growing phenomenon of the 21st century. *Philosophical Transactions of the Royal Society of London. Series B: Biological Sciences*, *357*(1420), 609–613.

Nordås, R., & Gleditsch, N. P. (2007). Climate change and conflict. *Political Geography*, *26*(6), 627–638.

Nordås, R., & Gleditsch, N. P. (2015). Climate change and conflict, In S. Hartard, & W. Liebert (Eds.), *Competition and conflicts on resource use* (pp. 21–38). Heidelburg: Springer Cham.

Null, S., & Risi, L. H. (2016, November). *Navigating complexity: Climate, migration, and conflict in a changing world*. Office of Conflict Management and

Mitigation Discussion Paper, Woodrow Wilson Center. https://www.wilsoncenter.org/sites/default/files/ecsp_navigating_complexity_web_0.pdf
Oliver-Smith, A. (2009). Climate change and population displacement: Disasters and diasporas in the twenty-first century. In S. A. Crate, & M. Nuttall, *Anthropology and climate change: From encounters to actions* (pp. 116–136). Walnut Creek, CA: Left Coast Press.
Paavola, J., & Adger, W. N. (2006). Fair adaptation to climate change. *Ecological Economics*, *56*(4), 594–609.
Paris, R. (2001). Human security: Paradigm shift or hot air? *International Security*, *26*(2), 87–102.
Perch-Nielsen, S. L., Bättig, M. B., & Imboden, D. (2008). Exploring the link between climate change and migration. *Climatic Change*, *91*(3–4), 375–393. https://doi.org/10.1007/s10584-008-9416-y
Piguet, E. (2008). *Climate change and forced migration*. New Issues in Refugee Research, Research Paper No. 153, Policy Development and Evaluation Service, United Nations High Commissioner for Refugees. Geneva: Switzerland. http://www.unhcr.org/en-au/research/working/47a316182/climate-change-forced-migration-etienne-piguet.html
Piguet, E., Pécoud, A., & De Guchteneire, P. (2011). Migration and climate change: An overview. *Refugee Survey Quarterly*, *30*(3), 1–23. https://doi.org/10.1093/rsq/hdr006
Raleigh, C. (2010). Political marginalization, climate change, and conflict in African Sahel States. *International Studies Review*, *12*(1), 69–86. https://doi.org/10.1111/j.1468-2486.2009.00913.x
Raleigh, C., & Jordan, L. (2010). Climate change and migration: Emerging patterns in the developing world. In R. Mearns, & A. Norton (Eds.), *The social dimensions of climate change: Equity and vulnerability in a warming world* (pp. 103–133). Washington, DC: World Bank,- New frontiers of social policy.
Raleigh, C., & Urdal, H. (2007). Climate change, environmental degradation and armed conflict. *Political Geography*, *26*(6), 674–694. https://doi.org/10.1016/j.polgeo.2007.06.005
Reuveny, R. (2007). Climate change-induced migration and violent conflict. *Political Geography*, *26*(6), 656–673. https://doi.org/10.1016/j.polgeo.2007.05.001
Reuveny, R. (2008). Ecomigration and violent conflict: Case studies and public policy implications. *Human Ecology*, *36*(1), 1–13. https://doi.org/10.1007/s10745-007-9142-5
Salehyan, I. (2008). From climate change to conflict? No consensus yet. *Journal of Peace Research*, *45*(3), 315–326. https://doi.org/10.1177%2F0022343308088812
Salehyan, I. (2014). Climate change and conflict: Making sense of disparate findings. *Political Geography*, *43*, 1–5, https://doi.org/10.1016/j.polgeo.2014.10.004
Scheepers, P., Gijsberts, M., & Coenders, M. (2002). Ethnic exclusionism in European countries. Public opposition to civil rights for legal migrants as a response to perceived ethnic threat. *European Sociological Review*, *18*(1), 17–34.
Scheffran, J. (2011). Security risks of climate change: Vulnerabilities, threats, conflicts and strategies. In H. G. Brauch, Ú. O. Spring, C. Mesjasz, J. Grin, P. Kameri-Mbote, B. Chourou, … J. Birkmann (Eds.), *Coping with global environmental change, disasters and security* (pp. 735–756). Berlin, Heidelberg: Springer.
Scheffran, J., Marmer, E., & Sow, P. (2012). Migration as a contribution to resilience and innovation in climate adaptation: Social networks and co-development in

Northwest Africa. *Applied Geography*, *33*, 119–127. https://doi.org/10.1016/j.apgeog.2011.10.002

Schleussner, C.-F., Donges, J. F., Donner, R. V., & Schellnhuber, H. J. (2016). Armed-conflict risks enhanced by climate-related disasters in ethnically fractionalized countries. *Proceedings of the National Academy of Sciences*, *113*(33), 9216–9221. https://doi.org/10.1073/pnas.1601611113

Siddiqui, T. (2010). *Impact of climate change: Migration as one of the adaptation strategies* (Working paper series no. 18). Dhaka: RMMRU. http://www.rmmru.org/newsite/wp-content/uploads/2013/07/workingpaper18.pdf

Siddiqui, T., & Billah, M. (2014). Adaptation to climate change in Bangladesh: Migration the missing link. In S. Vachani & J. Usmani (Eds.), *Adaptation to climate change in Asia* (pp. 117–141). Cheltenham: Edward Elgar. https://doi.org/10.4337/9781781954737

Stewart, F., Holdstock, D., & Jarquin, A. (2002). Root causes of violent conflict in developing countries/Commentary: Conflict—from causes to prevention? *British Medical Journal*, *324*(7333), 342–345.

Swain, A. (1996). Environmental migration and conflict dynamics: Focus on developing regions. *Third World Quarterly*, *17*(5), 959–974.

Tacoli, C. (2009). Crisis or adaptation? Migration and climate change in a context of high mobility. *Environment and Urbanization*, *21*(2), 513–525. https://doi.org/10.1177%2F0956247809342182

Tacoli, C., McGranahan, G., & Satterthwaite, D. (2015). *Urbanisation, rural-urban migration and urban poverty*. Working Paper, March 2015. London: International Institute for Environment and Development (IIED). http://pubs.iied.org/pdfs/10725IIED.pdf

Theisen, O. M., Gleditsch, N. P., & Buhaug, H. (2013). Is climate change a driver of armed conflict? *Climatic Change*, *117*(3), 613–625. https://doi.org/10.1007/s10584-012-0649-4

UNHCR. 2002. Environmental migrants and refugess. *Refugees*. No. 127. http://www.unhcr.org./pub/PUBL/3d3fecb24.pdf

Urdal, H. (2005). People vs. Malthus: Population pressure, environmental degradation, and armed conflict revisited. *Journal of Peace Research*, *42*(4), 417–434.

Van Hear, N., Bakewell, O., & Long, K. (2018). Push-pull plus: Reconsidering the drivers of migration. *Journal of Ethnic and Migration Studies*, *44*(6), 927–944. https://doi.org/10.1080/1369183X.2017.1384135

Verhoeven, H. (2011). Climate change, conflict and development in Sudan: Global neo-Malthusian narratives and local power struggles. *Development and Change*, *42*(3), 679–707. https://doi.org/10.1111/j.1467-7660.2011.01707.x

Walsham, M. (2010). *Assessing the evidence: Environment, climate change and migration in Bangladesh*. Geneva: International Organization for Migration (IOM), Regional Office for South Asia. https://www.iom.int/jahia/webdav/site/myjahiasite/shared/shared/mainsite/events/docs/Assessing_the_Evidence_Bangaldesh.pdf

Warner, K. (2010). Global environmental change and migration: Governance challenges. *Global Environmental Change*, *20*(3), 402–413. https://doi.org/10.1016/j.gloenvcha.2009.12.001

Warner, K., Hamza, M., Oliver-Smith, A., Renaud, F., & Julca, A. (2010). Climate change, environmental degradation and migration. *Natural Hazards*, *55*(3), 689–715.

Webber, M., & Barnett, J. (2010). *Accommodating migration to promote adaptation to climate change*. Policy Research Working Paper. Washington, DC: The World Bank. https://doi.org/10.1596/1813-9450-5270

Weiner, M. (1992). Security, stability, and international migration. *International Security*, *17*(3), 91–126.

World Bank. (2017). Migration and remittances: Recent developments and outlook. *Migration and Development Brief*, no-27. Washington, DC: The World Bank. http://pubdocs.worldbank.org/en/992371492706371662/MigrationandDevelopmentBrief27.pdf

Zaman, M., & Wiest, R. (1991). Riverbank erosion and population resettlement in Bangladesh. *Practicing Anthropology*, *13*(3), 29–33.

3 Climate change, migration, and conflict issues in Bangladesh

Introduction

This chapter presents the context of the study of Bangladesh, presupposing that environment and climate change–induced migration from the lowland and rural areas to the urban areas and the CHT has connections with the conflict and violence in the region. Bangladesh is already regarded as one of the most climate change affected countries in the world (IPCC, 2014; Maplecroft, 2011). The location of the country, its riverine character as a floodplain, and its lower riparian position in the Ganges basin have positioned the country as a victim of floods, cyclones, coastal erosion, and drought (Agrawala et al., 2003; Brouwer et al., 2007; Mirza & Ahmad, 2005). The socioeconomic conditions, overpopulation, uneven development, and unplanned urbanization have also made the country vulnerable to climate change risks (Shamsuddoha & Chowdhury, 2007). Climate change issues, both slow processes (sea-level rise, drought, and desertification) and sudden processes (floods, cyclones, and riverbank erosion), are affecting the environment, livelihood, and overall development of Bangladesh. One of the most crucial impacts of climatic events is human displacement and migration. Different studies have offered estimates of the likely displacement of people in Bangladesh due to environment and climatic events, such as "13.5 million people by 2050" (Rigaud et al., 2018: 144) and "around 50 million people by 2050" (Shamsuddoha & Chowdhury, 2007: 763). Although the majority of internal migration is to cities such as Dhaka, Khulna, and Chittagong (Afsar, 2003; Bryan, Chowdhury, & Mobarak, 2014; Siddiqui, 2005), a significant number of migrants have settled in the Chittagong Hill Tracts (CHT) with the help of successive governments. A number of studies have argued that successive governments at various times have used resettlement as a way to reduce population pressure and avoid environmental vulnerabilities (Lee, 1997; Reuveny, 2007; Suhrke, 1997).

The CHT is a distinct and unique area in Bangladesh which has a history of occupation by foreign rulers such as the Arakan and British colonial powers (Van Schendel, Mey, & Dewan, 2000; Mohsin, 1997). Thus, immediately after the independence of Bangladesh, leaders of the CHT demanded autonomy,

DOI: 10.4324/9781003430629-3

retention of the 1900 CHT manual (Regulation of 1900 Act), the constitutional recognition of their separate identity, and a ban on the influx of non-tribal people into the CHT (Mohsin, 1997: 57–58; Levene, 1999). However, this was denied by successive governments. After 1975 the region experienced ethnopolitical conflict between the Bangladesh army and Shanti Bahini. Although the signing of a peace accord in 1997 between the combatants ended the armed conflict, a broad form of social conflict between the Bengali settlers and tribal people continues to affect every sphere of life in the CHT.

This chapter first presents the demographic scenario of Bangladesh in order to develop the idea of how the size of the population has increased over time. This assists in understanding the relationship between population pressure and migration. Second, this chapter discusses the state of climatic events and how climatic events influence internal displacement and migration in the lowland and rural areas in Bangladesh. The chapter then describes the migration of Bengali people to the CHT and the demographic transformation this has produced in the region. This is followed by a discussion of the existing research in a broad context of the CHT conflict and how the migration is connected to it.

The demographic scenario in Bangladesh

Demography is an important factor in the overall socioeconomic and development issues in Bangladesh. The population size in 1974 was 76 million, and by 2011, when the last population census was conducted, it had almost doubled (Figure 3.1). The population in 2018 was more than 166 million, and the population density is more than 1,250 per square kilometer (World Population Review, 2018). This large population size has both positive and negative implications. From a positive standpoint, a growing population entails a large workforce able to contribute to the economy through exploring production opportunities and exporting possibilities. In the last two decades,

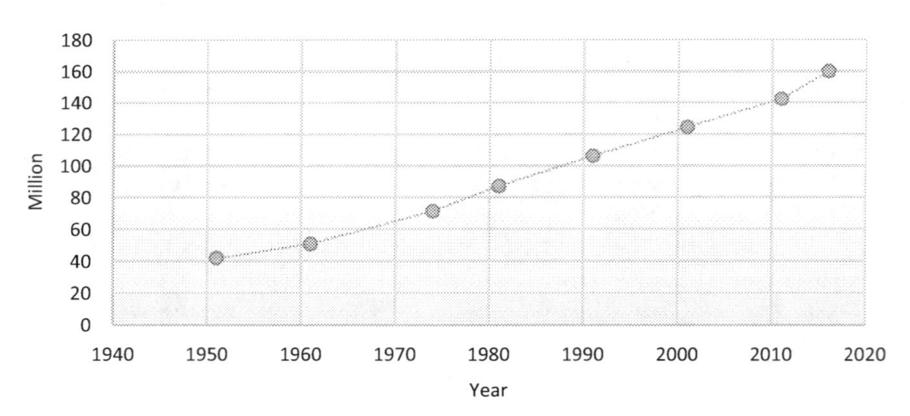

Figure 3.1 Population trends in Bangladesh (1940–2016)

Sources: Developed by the researcher based on population census in Bangladesh.

the labour force, which was exported to the Middle East and developed countries, has earned billions of dollars and contributed significantly to remittances returned to Bangladesh (Chowdhury, 2011; Siddiqui, 2005).

However, overpopulation is significantly affecting the overall development of Bangladesh. The population pressure causes unemployment, pressure on resources, and land scarcity and increases the gap between rural and urban areas. The significant impact of the increasing population is the shrinking of per capita land. It is a fact that Bangladesh is one of the most land-scarce countries in the world with only 12.5 decimals[1] per capita (Quasem 2011: 59). The high population growth rate has caused land fragmentation, which hinders the introduction of modern agricultural technologies and leads to reduced productivity (Rahman & Rahman, 2009). Population pressure, poverty, and declining arable land are key factors in a growing number of people migrating from rural areas to other places within and beyond the country for income opportunities and a better future.

Urbanization and development scenario in Bangladesh

With the partition of British India in 1947, increasing migration of Muslims has contributed to a near 44.63 percent increase in urban population in Bangladesh between 1951 and 1961 (Khan, 1982). Primary reasons included employment for nearly 80 percent of these migrants, with other reasons including better living amenities, education and medical benefits, and better social standings (Islam, Khan, & Khan, 1975). The 1981 official census shows data indicating this trend still going strong with Dhaka having over 38 percent and Chittagong having over 31 percent of total urban population in the country due to internal migration (Kemper, 1989). The rapid growth of the 1980s has decreased to a still high annual exponential rate of growth for urban population in the country (3.25% in the year 2001), as the following Table 3.1 shows (Khan, 2008).

The annual growth rate has stayed strong and rising throughout the years, and the latest statistical data can be accumulated to show the trend of urban population growth in the country, compared to the rural population growth rates (Figure 3.2).

As of 2018, the urban populations post–21st century largely consist of people who have migrated from rural settings, with three out of every five migrant arriving from rural areas in Dhaka alone (Amin, 2018). The generally low levels of skill and lower levels of literacy of these rural migrants have led to them being involved with low-paying wages, notably in the garments and transportation sector. Their living conditions are also severely lacking, leading to sprawling urban slums in the urban areas of the country (Khan et al., 2018). With recent years, the push factor of a large portion of these rural migrants has been the climatic events in their respective rural areas, often more so than the existing benefits of urban settings. The climatic events have led to deteriorating opportunities in rural areas, with families losing

Table 3.1 Growth of urban population in Bangladesh, 1951–2001

Census year	*Total national population (million)*	*Annual growth rate of national population (%)*	*Total urban population (million)*	*Urban population as percentage of total population (i.e. level of urbanization)*	*Decadal increase of urban population (%)*	*Annual exponential growth rate of urban population (%)*
1951	44.17	0.50	1.83	4.34	18.38	1.58
1961	55.22	2.26	2.64	5.19	45.11	3.72
1974	76.37	2.48	6.00	8.87	137.57	6.62
1981	89.91	2.32	13.56	15.54	110.68	10.03
1991	111.45	2.17	22.45	20.15	69.75	5.43
2001	123.10	1.47	28.81	23.40	27.38	3.25

Source: (Khan, 2008).

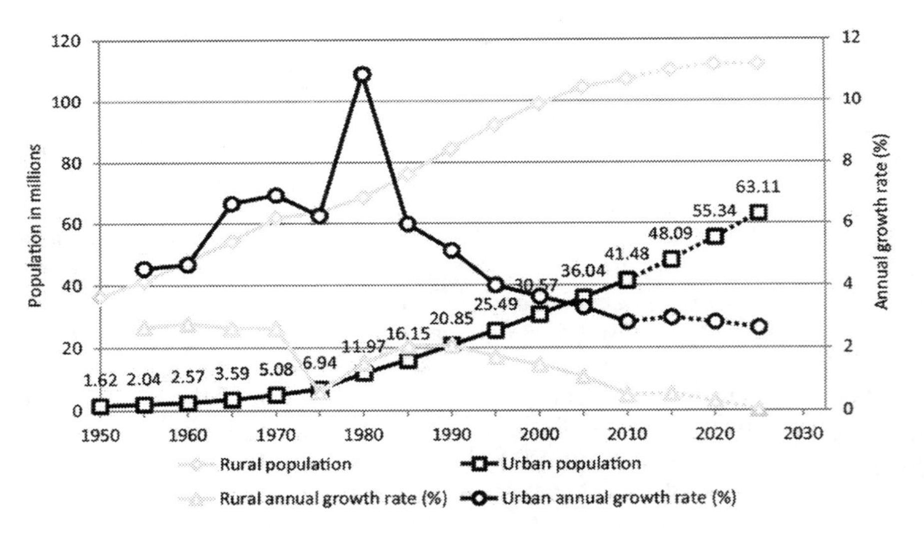

Figure 3.2 Trend of urban population growth in Bangladesh from 1950 to 2030
Source: Ahmad (2015).

property and livestock to cyclones, droughts, floods, and river erosion (Ahsan, 2019). This has added to the already existing plethora of incentives for these people to migrate to towns and cities.

Climate change issues

As one of the largest deltas in the world (see Figure 3.3), Bangladesh is highly vulnerable to climatic events. This vulnerability is due to its low elevation above sea level, downstream location, and floodplain dominance.

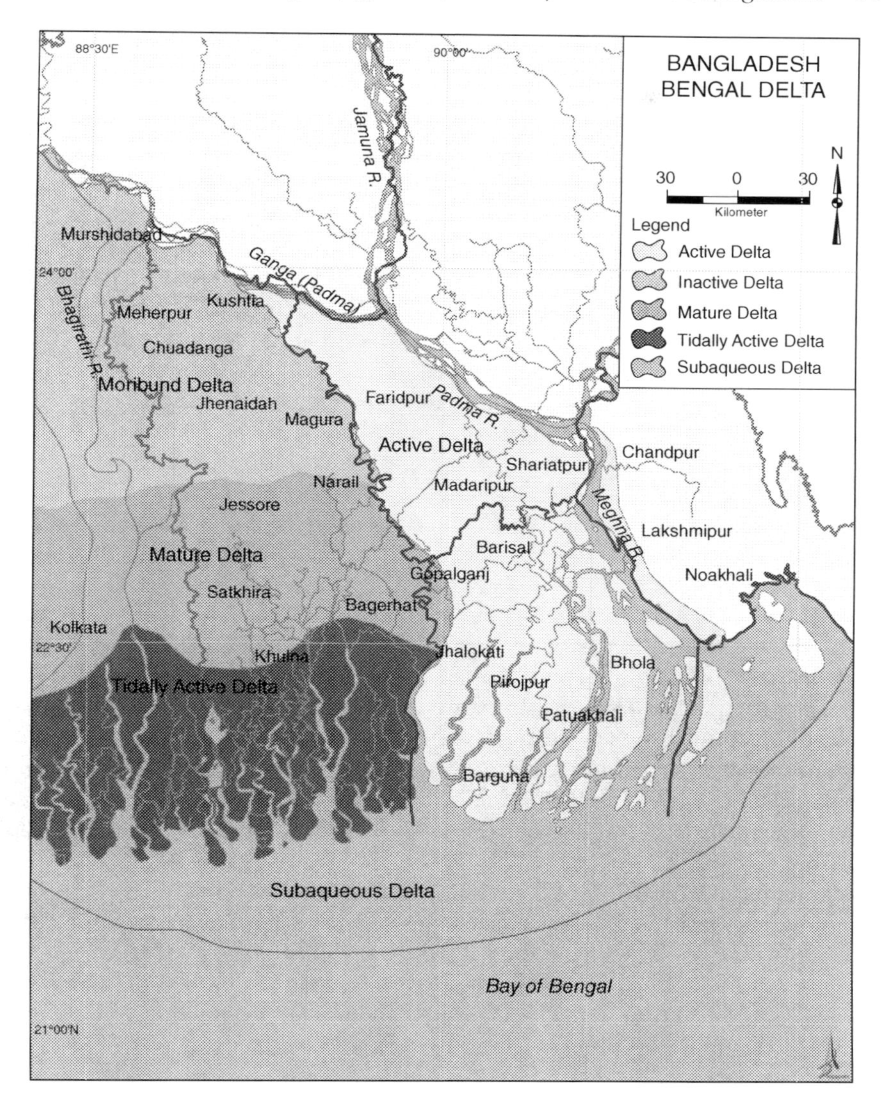

Figure 3.3 Delta areas in Bangladesh

Source: Banglapedia: The National Encyclopaedia of Bangladesh (2012).

About climate change–induced disasters, Quencez (2011, 59) notes that "Bangladesh is already considered to have the highest risk of flooding out of 162 countries; the third highest risk of tsunamis out of 76 countries; the sixth highest in terms of cyclone threat of 89 countries; the 63rd out of 184 for drought and 35th out of 162 prone to landslides." The following sections provide a description of the major climate change issues in Bangladesh.

Sudden climatic events: floods and cyclones

Sudden climatic events such as floods, cyclones, and river erosion are major concerns for Bangladesh. These three events occur every year and cause detrimental impacts on the population, ecosystems, and overall development. Floods are an inescapable fact for many people in Bangladesh (Walsham, 2010). Every year Bangladesh experiences floods which inundate at least 30 percent of the landmass. In extreme cases, 70 percent of the country is under flood (Mirza, Warrick, & Ericksen 2003: 287). Over the last two decades, the country has experienced six devastating floods, along with annual flooding, which have caused the displacement of 30–45 million people from their homes (Walsham, 2010: 9). The severe floods in 2007 inundated a 32,000 square kilometer area affecting 16 million people and 3 million households (Walsham, 2010: 10). Figure 3.4 shows that most parts of the country are affected by floods.

Moreover, floods cause water logging in low-lying areas and make the living places unliveable. People in some locations such as Khulna, Barisal, Satkhira, Coxes Bazar, and Patuakhali district suffer most from the water logging.

Other sudden onset climatic events are cyclones and storms. The country is faced with different forms of cyclones and storms which are significant security threats to human lives and habitat. Figure 3.5 shows the adjacent districts of the Chittagong and Khulna divisions to the Bay of Bengal are highly affected by cyclones and storms that cause havoc and human casualties.

The most important effect of these cyclones and storm surges is population displacement. In 1970–1998, the country suffered from 170 large-scale cyclones and storm surges, many of them with devastating consequences (Penning-Rowsell, Sultana, & Thompson, 2013). For example, "two major cyclones in 1970 and 1991 killed 500,000 and 140,000 persons respectively" (Penning-Rowsell et al., 2013: S45). In 2007 the cyclone Sidr[2] struck the coast of Bangladesh, causing the loss of over 10,000 lives and shattering the livelihoods and living options of over 30 million people. In the following year (2008) another strong and devastating cyclone Nargis[3] hit the country and killed more than 100,000 people in Bangladesh and Myanmar. On 25 May 2009, cyclone Aila hit the coasts of Bangladesh and India and killed around 1000 people and dislocated the lives of several thousand people.

Slow onset climatic events: sea-level rise and drought

In the category of slow-onset processes, sea-level rise causes displacement and vulnerabilities to the people of coastal areas. Half of Bangladesh is situated less than 5 meters above sea level, and one-third is less than three metres (Myers & Kent, 1995: 117). Sea-level rise causes saline intrusion and floods in many coastal areas in Bangladesh (Walsham, 2010). It is predicted that a 1 meter sea-level rise would affect 22,000 square kilometers of the deltaic area inhabited by more than 17–18 million poor people in Bangladesh (Bose,

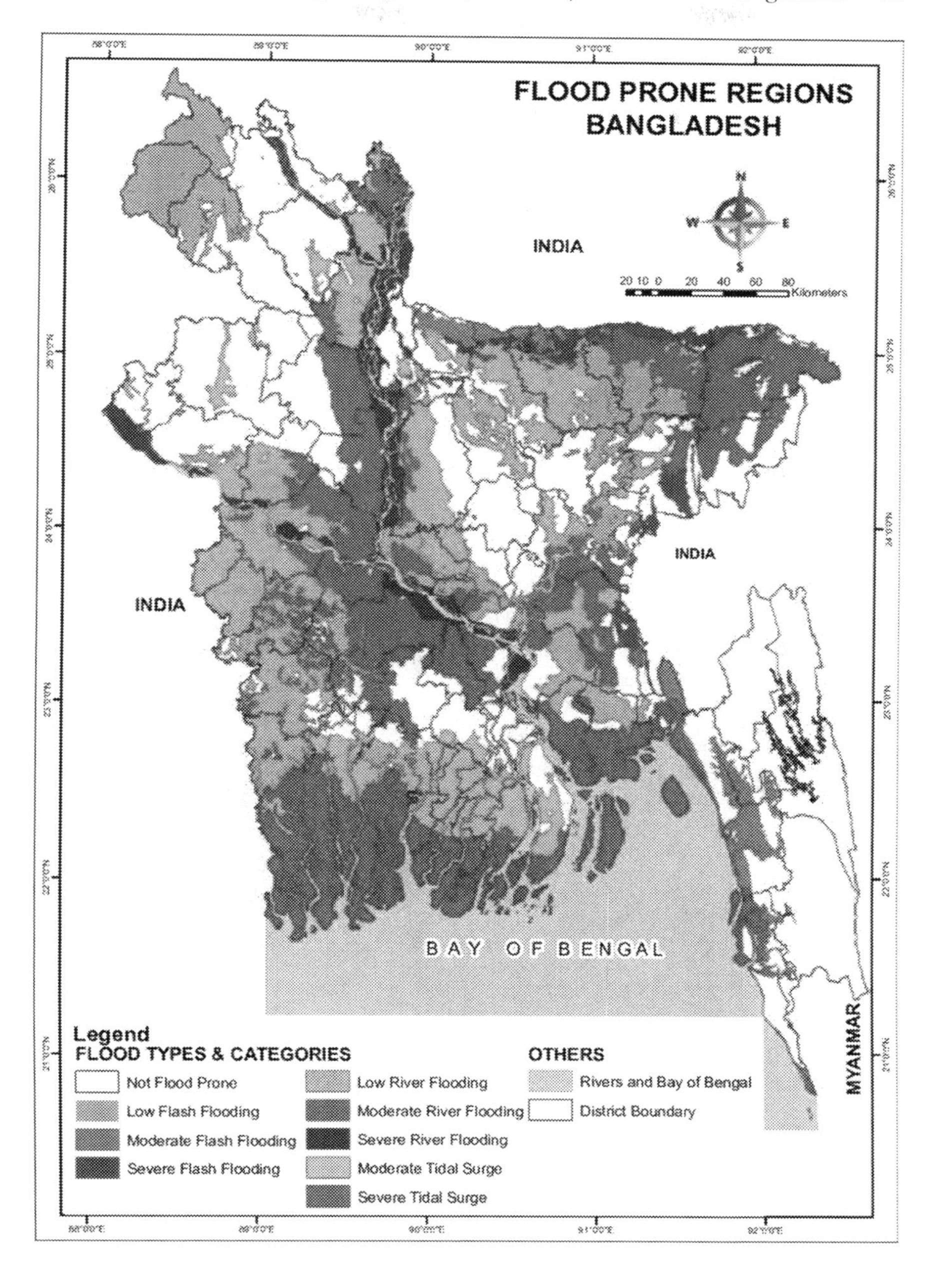

Figure 3.4 Flood-prone regions in Bangladesh

Source: Climate Change Cell, 2006.

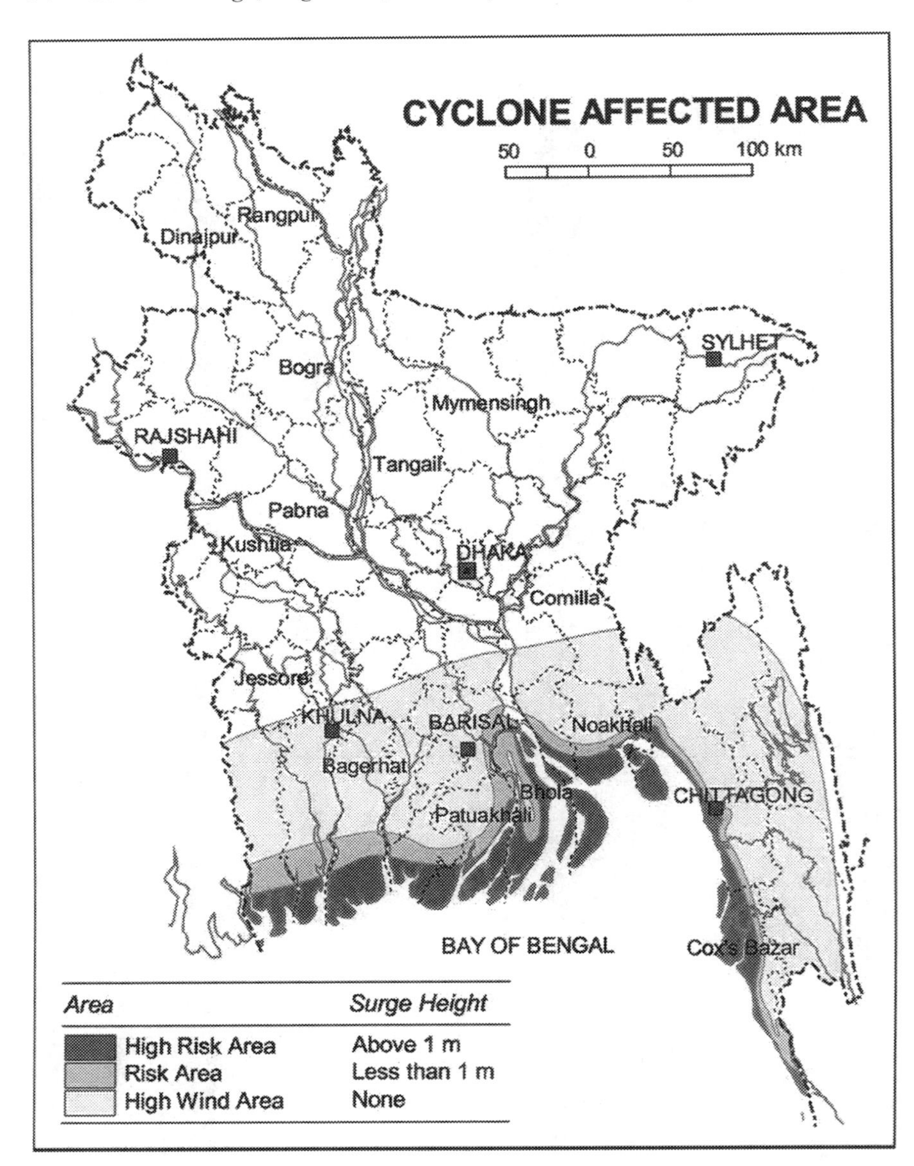

Figure 3.5 Cyclone-prone areas in Bangladesh

Source: Climate Change cell, 2006; Bangladesh Space Research and Remote Sensing. SPARRSO.

2013: 64). Sea-level rise in the coastal zone of Bangladesh has already been observed to cause erosion and inundation of the land and salinization of soil and water, and cause flooding from the storm surge. Figure 3.6 shows the extent of the affected areas by the sea-level rise.

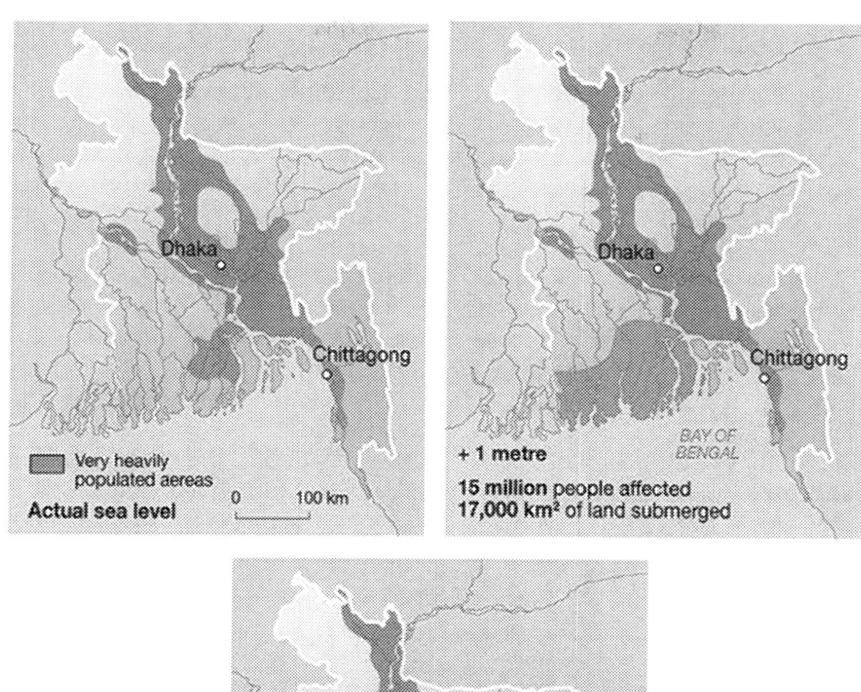

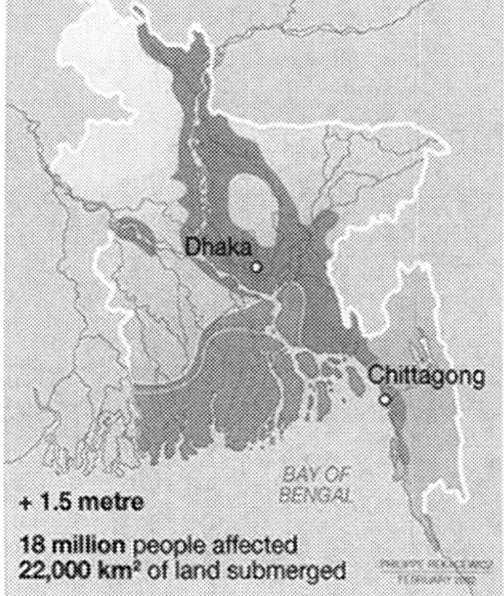

Figure 3.6 Extent of the affected areas by the sea-level rise

Source: University Corporation for Atmospheric Research (UCAR) (2018). Dacca University: Intergovernemntal Pannel on Climate Change (IPCC).

In Bangladesh the coastal region constitutes 20 percent of total land area and over 30 percent of the cultivable land (Minar, Hossain, & Shamsuddin, 2013: 114) where 63 million people live, the third highest among countries having coastal people (Neumann et al., 2015). Based on the population growth rate of 1.4 percent per year, a study has suggested that 6.8 million people would be at risk by 2025 and 12.7 million people by 2050 under the prevailing climate change conditions (Karim & Mimura, 2008: 497).

Meanwhile, the northern districts of Bangladesh are increasingly affected by droughts. Rising temperature, lack of water in the dry season and drying rivers are responsible for the desertification processes in Bangladesh. Scientists predict that the climate in Bangladesh may become warmer in future. The first IPCC (1990) report projected that Bangladesh will be 0.5°C to 2°C warmer than today by the year 2030 (Ahmed, 2006: 5). Desertification is already happening in some parts of Bangladesh. The northwestern part of Bangladesh is a drought-prone area where rainfall is less than half the national average (1,240 mm/year). The area is already experiencing "increasing temperature (0.05 °C/year) and a decline in the length of the monsoon" (Black, Kniveton, et al., 2013b: 46). Currently, the country is exposed to climate change effects, as well as water diversion from upstream international rivers by neighbouring states. Both climate change–induced rainfall shortage and shortage of water in the rivers are causing drought in parts of Bangladesh. A study report claims that "around 213 upozilas (administrative units consist of some villages) are drought affected in Bangladesh and increasing climate change may intensify the scenario in future (CDMP II, 2013: 77).

Event associated to climatic change

Riverbank erosion (Figure 3.7) is not a direct climatic event but results from floods, storm surges, and cyclones. Human intervention also contributes to weakening riverbanks when more people live on the riverbank or cut land from the bank for other purposes. Based on statistics from the Bangladesh

Figure 3.7 Riverbank erosion in Kurigram district

Source: Green Watch (2016).

Water Development Board, Pender (2008: 34) reported that "1,200 km of riverbank has been actively eroded and more than 500 km faces severe problems related to erosion". Riverbank erosion destroys the landmass of the riverside and washes away houses, crops, trees, and other means of livelihood of people living on the sides of the rivers. In some cases, an entire village is destroyed and disappears into the river water.

In such situations, people are forced to migrate to the adjacent areas for temporary shelter and livelihood. In some areas, riverbank erosion occurs several times, and people experience multiple migrations in their lifetime (Arsenault, Azam, & Ahmad, 2015). The cumulative impact is significant, with estimates that "riverbank erosion displaces 50,000 to 200,000 people in Bangladesh every year" (Mehedi, Nag, & Farhana, 2010: 5). Another study claims that 60 percent of the residents have been displaced in their lifetime in some of the worst affected areas in Bangladesh (Raleigh & Jordan, 2010: 115). Cities such as Chandpur, Rajshahi, Sirajgong, Gaibandha, Rajshahi, Bikrampur, Shariatpur, and Faridpur are heavily exposed to riverbank erosion (see Figure 3.8).

Affected people try to reestablish their houses once new *char*[4] is formed near their original land; however, at least 10–20 percent of displaced people are not able to return to the newly formed *char* due to their poverty, the uncertainty of making a living, and fear of being a victim of flooding again (Raleigh & Jordan, 2010: 24). Increasing erosion has had significant impacts on coastal land. Almost all islands are shrinking at an increasing rate due to the continuous erosion caused by sea-level rise and storm surges. For instance, in the last 40 years Bhola Island has shrunk from 6,400 square kilometers to 3,400 square kilometers, about 40 percent (Figure 3.8). Significant erosion has occurred in the north of Hatia, northeast of Bhola and southeast of Ramgati islands (Brammer, 2014: 53). A study (Shamsuddoha & Chowdhury, 2007) showed that around 65 percent of the landmass of these islands has been eroded within the last 100 years. But the population in coastal areas is increasing at the same rapid pace over the entire country (Figure 3.9).

The coastal population will increase to be 44 million if the current population growth rate remains unchanged (Shamsuddoha & Chowdhury, 2007: 6). Climatic events in coastal regions affect the people most and displace them from their place. Displaced people migrate to the remaining parts of the islands and coastal regions that consequently have an increase in population density.

Population movement due to climatic events in Bangladesh

Climatic events have far-reaching impacts on population displacement in Bangladesh. Although there is no official data on climate change induced migration, various sources such as NGO reports and researchers have given estimates of its extent (Hassani-Mahmooei & Parris, 2012; Ahsan, Kellett, & Karuppannan, 2014; Davis et al., 2018). However, most figures are based on

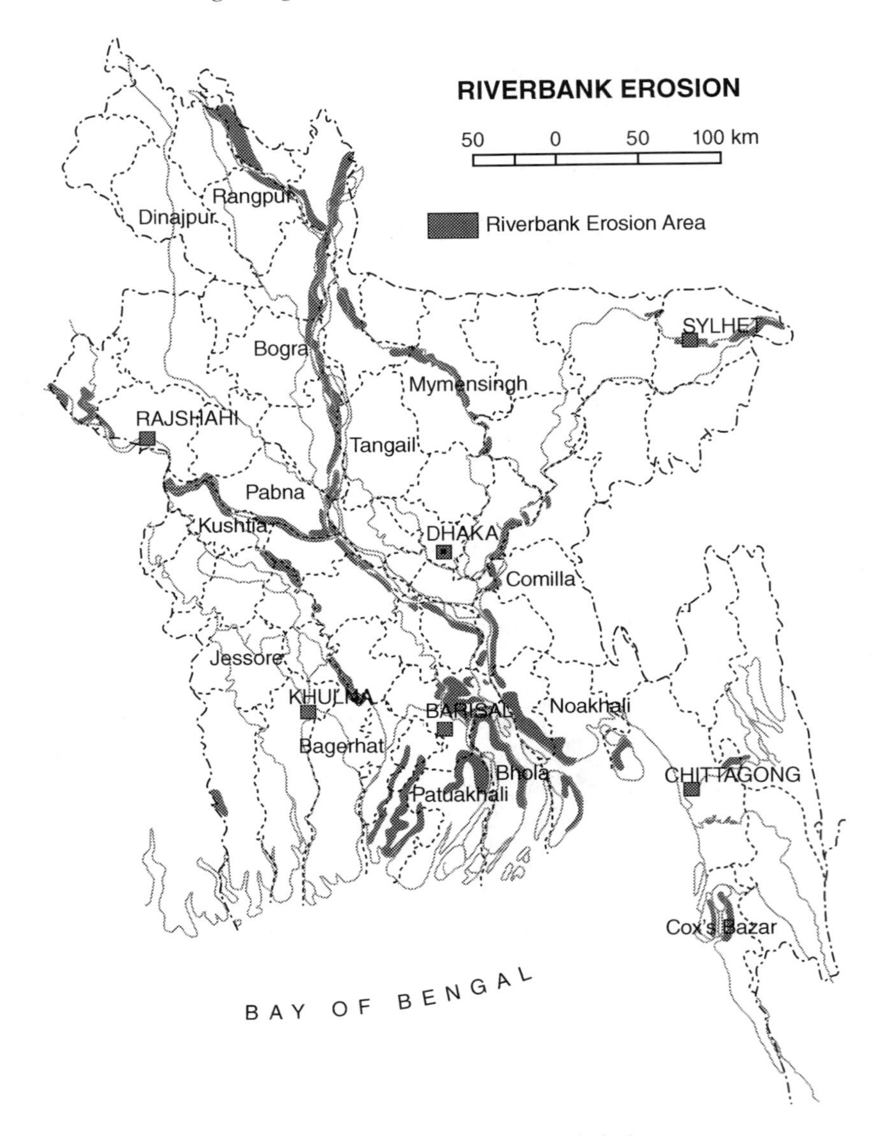

Figure 3.8 Areas affected by riverbank erosion in Bangladesh

Source: Bangladesh Water and Development Board.

sporadic data. The population census report in 2001 projected that rural and urban populations would be almost equal due to the increasing internal migration. Several factors are responsible for urbanization, but the most evident cause is the migration of rural people to urban centers (Ahsan, Kellett, & Karuppannan, 2016).

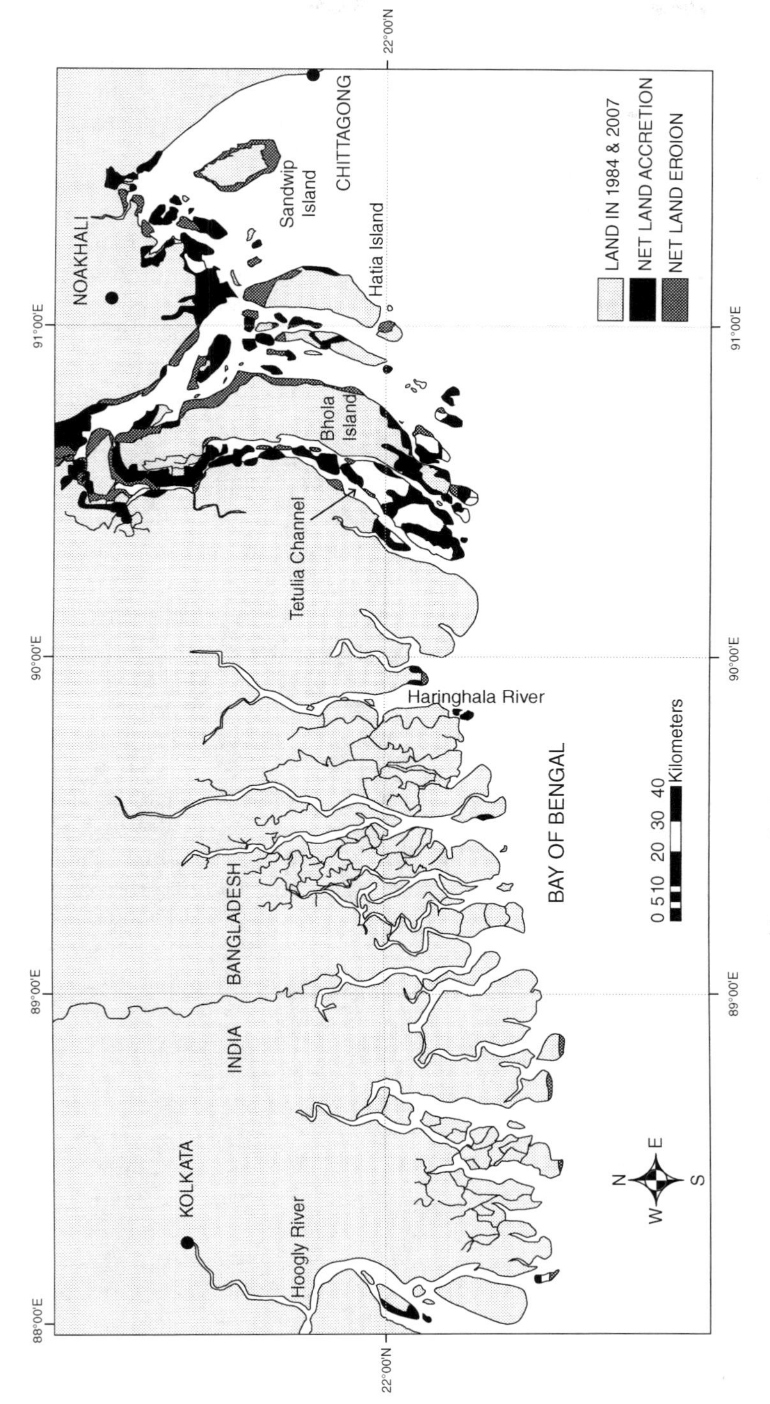

Figure 3.9 Land erosion in Bhola, Hatia, Sandwip, and other islands in Bangladesh

Source: Google Earth image.

These climate and environmental events force people to migrate temporarily or permanently by destroying agricultural land and livelihood options. In many locations, floods and cyclones indiscriminately destroy all possible sources of livelihood and shelter. Having lost everything, people migrate as a last resort (Penning-Rowsell et al., 2013). A study conducted in the four districts in the east and west parts of the country shows that "out of 595 rural households, 168 (28%) households indicate that they have at least one migrant; 79 percent of these 168 households have only one migrant, others have more than one" (Rayhan & Grote, 2007: 88). The pattern of migration is rural to rural and rural to urban. The study also shows that around 6 percent of the respondents migrated to adjacent areas (Rayhan & Grote, 2007: 89). A study by Joarder and Miller (2013) finds that climate affected people are more likely to decide to permanently migrate when they have lost their household, arable land and sources of livelihood from drought, riverbank erosion, and floods.

However, some researchers are critical and skeptical about the impact of climate change on migration decisions. Findlay and Geddes (2011) found less connection between climate change and migration in Bangladesh than other researchers. They find that people affected by climatic events may not migrate, but rather attempt to remain in their place and try to adapt to the changing environment (Findlay & Geddes, 2011: 146–7). In some cases, people migrate to the cities temporarily to earn more income and return to their homes after a few months (Call et al., 2017; Lein, 2000). This is a coping strategy of climate change induced people in Bangladesh to reduce their vulnerabilities (Siddiqui & Billah, 2014). Another study conducted in north Bengal in Bangladesh shows that climate change is not the sole cause of migration. Social inequality, poverty, and food insecurity play a role in people migration from rural areas (Etzold et al., 2014; Quader, Khan, & Kervyn, 2017). Chronic poverty mostly caused by the drought in the north Bengal of Bangladesh (Mazumder & Wencong, 2012) acts as a leading cause of migration to cities.

Migration due to climate change depends on a complex set of reasons. Poverty, insufficient infrastructure, and lack of government assistance during disasters may combine in forcing people to migrate (Mallick & Vogt, 2012). Labour market disruption, inflation, and sudden price hikes in the cost of food lead some people to migrate from their villages (Rayhan and Grote, 2007). These are economic push factors that force people to move to the cities for their livelihood. In Bangladesh, poverty and landlessness constitute crucial factors for internal migration. Both sudden and slow onset of climatic events cause poverty and landlessness by destroying the infrastructure, land, bridges, and embankment of the rivers. In the lowland and riverside areas, floods and riverbank erosion cause the landlessness. Many landless people migrate to cities when they struggle to make a living in their rural place of origin. It is argued that "half of all poor migrants to Dhaka are landless labourers, while three-quarters of women and two-thirds of men working in textile factories have been found to be functionally landless" (Black et al., 2008: 30).

Furthermore, sudden climatic events, such as floods and cyclones, result in crop failure and contribute to internal migration. Based on longitudinal survey data, Gray and Mueller (2012) argue that "floods have modest effects on population displacement, but crop failure unrelated to the floods has a strong correlation to population displacement in Bangladesh" (Gray & Mueller, 2012: 1). Crop failure is also connected to the cycle of poverty and indebtedness. Periodic crop failure intensifies the poverty situation that sometimes pushes farmers to obtain loans from NGOs to support their farming activities and small businesses. However, further crop failure and loss of business prevents them from repaying the loan. Some poor people living in urban slums have moved to avoid money lenders (Findlay & Geddes, 2011).

Poor people in rural Bangladesh have a long history of seasonal migration to work in the fields and earn extra income; however, increasing climatic events and associated stresses in the rural and coastal areas contribute to increased temporary seasonal migration. Climate change affected families sometimes prefer to send at least one family member to the city or overseas for income diversification and improved livelihood (Afsar, 2003; Siddiqui & Billah, 2014). All of this is contributing to the urbanization of Bangladesh and the growth of its capital city, Dhaka. Black, Kniveton, et al. (2013b: 44 citing Afsar, 2003) claim that "the net migration from rural to urban areas in Bangladesh has been increased dramatically from 1.2 to 16.4 per thousand between 1984 and 1998, compared to an increase from 1.5 to 4 per thousand of rural-rural migration during the same period." Siddiqui and Billah (2014) report that since the 1990s about 5 to 10 percent of people who belong to the middle class have migrated to cities for a better life from the Shatkhira region of Bangladesh, one of many areas that are highly affected by floods, coastal erosion, and cyclones (Siddiqui & Billah, 2014: 131). As this literature suggests, poor people affected by climatic events tend to engage in seasonal migration to the cities for livelihood and income. People from the lowland and rural areas move to more fertile regions for seasonal work opportunities, which is a common phenomenon in Bangladesh, but many seasonal migrants eventually make more permanent migration decisions. For example, many people from the lowland areas of Noakhali, Chandpur, and Cumilla have settled in the elevated parts of the southern districts (e.g.,, Jessore, Khulna, and Kushtia) in Bangladesh. Some better-off people also migrate to the cities from climate affected areas as they have resources to leave their place of origin and settle in a new place.

Resilience, adaptation, and mitigation support also influence the migration decisions of the people affected by floods, cyclones, and drought. Several studies suggest that affected people move an average of 2 miles from their homes, believing that they would be able to settle on land that will be reclaimed somewhere close to their home (Mutton & Haque, 2004; Lein, 2000: 124). However, the formation of new land (mostly called char) can take a long time and in most cases, influential people take control of newly formed land. Thus, displaced people rarely have an opportunity to return to resettle near their houses (Zaman, 1989), and may end up migrating to the city and adopt any sort of work for their shelter and livelihood. However, research also finds that climate affected people did not leave their homes when the government and nongovernment organizations distributed relief and provided support to adapt to the environment (Paul 2005; Paul & Routray, 2010). Consequently, internal migration of the rural poor people depends on the availability of livelihood. When government support and assistance is lacking, climate affected people may permanently abandon their homes (Kartiki, 2011).

In summary, it can be argued that climatic events have direct and indirect effects that negatively impact the livelihood and living conditions of millions of people and cause internal migration in Bangladesh. However, there is no comprehensive picture about internal migration, and to what extent it is caused by climate change. One reason is that the GoB has neglected internal migration in Bangladesh. Nevertheless, the existing research also argues that internal migration is seen as a survival strategy in Bangladesh. Many poor and landless people being affected by floods, cyclones, coastal erosion, and drought migrate to urban areas in search of livelihood and shelter.

Migration and demographic change in the urban areas and the CHT

The previous section argues that climate change displaces people who temporarily or permanently migrate to urban places. However, some studies argue that climate change–induced migrants and poor people have also migrated to the CHT region with the support of government settlement schemes, as well as of their own accord (Hafiz & Islam, 1993; Lee, 2001; Reuveny, 2007). It is argued that migration in Bangladesh is a natural process and people from the rural and coastal areas come to the urban areas for their livelihood and shelter. This in-migration of people has changed the structure of the urban areas in terms of population, infrastructure, services, and governance issues.

Climate change induced migration has been more and more prominent in the capital Dhaka in recent years. With nearly 2,000 new people settling there every day, the city now has a vast population of slum-dwellers, who have moved due to mainly events induced by climate change (Castellano, Dolšak, & Prakash 2021). It has been estimated that around 300,000 to 400,000 new migrants come to Dhaka from different parts of the country every year and primarily take residence in more than 5,000 slums across the Dhaka city. The number of slums has also nearly doubled in recent times across the metropolitan areas of Dhaka. The International Organization for Migration (IOM) has estimated that around 70 percent of these migrants in recent years have arrived due to feeling a climatic event of some form (Amjad, 2021). Among the manifold climatic events responsible for their migration, river erosion has been seen as the most prominent climatic event that has forced these climate migrants to move to Dhaka (Figure 3.10).

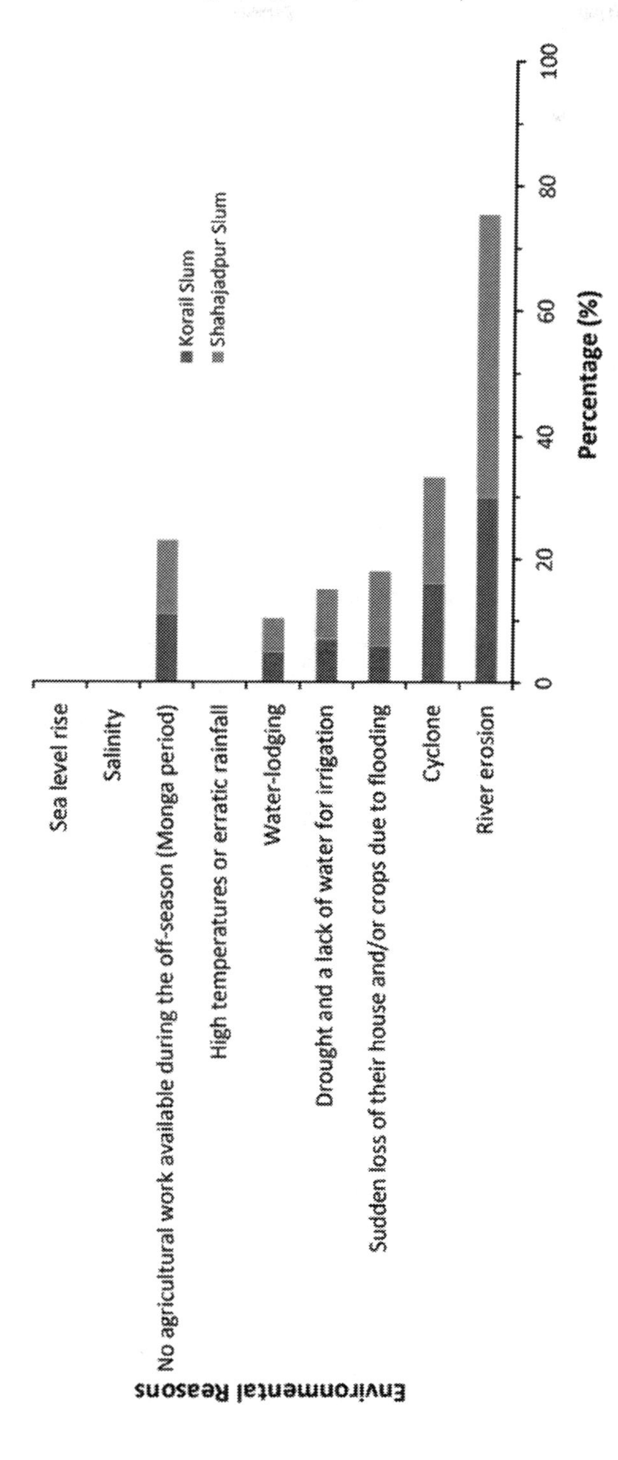

Figure 3.10 Environmental reasons for migrating

Source: Amjad (2021).

Figure 3.10, from a survey done of the two largest slums in Dhaka – Korail and Shahjadpur – demonstrates the distribution of climate change induced migration in Dhaka among various climatic events.

In 2008, a third of the population in the metropolitan areas of Dhaka was composed of slum-dwellers, which was around 37 percent (about 3.3 million) of a population of about 9.1 million people (Streatfield & Karar, 2008). Living in unhealthy conditions, these slum-dwellers lack basic rights of tenure and have issues related to health and sanitation deficits (Ahmed & Meenar, 2018). Lacking enough income and education, they often suffer from common diseases and receive little to no proper medical aid, leaving odds to self-care and quacks, aside from improvised basic medicinal remedies such as paracetamol. Only about a quarter of them are able to consult professional doctors, although not as frequent as needed (Mannan, 2018).

The number of the Bengali population in the CHT has increased gradually through the policy of different rulers from the British to Bangladesh period as a way of controlling the CHT region. Although the British colonial rulers introduced the 1900 Regulating Act to ensure the separate and distinct characteristics of the CHT, they introduced the cultivation of cotton, which eventually turned the hill area into cotton fields (*Karpash Mahali*) instead of the traditional farming system of the tribal people (for example, jhum cultivation) (Chatterjee, 1987; Tripura, 2008; Uddin, 2010). This introduction of cultivating cotton in the CHT enabled the non-hill Bengali people to work in the field (Mohsin 1997: 79). This is often cited as the beginning of a long process of Bengali migration to the CHT.

After the colonial rulers left, the Pakistan government (1947–1971) adopted twin policies in the CHT that fundamentally changed the region's social, cultural, and economic character. The government of Bangladesh pursued the settlement of the Bengali people to the CHT area and implemented development projects that displaced many of the tribal people from their land (Chakma, 2010; Levene, 1999; Shelly, 1992). The Pakistani government abolished the special status of the CHT and adopted an implicit policy of Islamization through the migration of Bengali Muslim people to the tribal-dominated area. As Pakistan emerged as a new state based on religious identity, the new government gave more emphasis to the consolidation of Islamic values in every corner of the country.[5] To bring the region under more direct control and integrate it into the nation-building project, the government pursued development projects which opened up the path for the Bengali Muslims people to work in the paper mills and hydro projects (Kharat, 2003; Mohsin, 1997). The permanent settlement of Bengalis was further assisted by the Pakistan government changing legislation (Amendment of Rule 34) to enable non-hill people to own property, provided that they had been living in the CHT for 15 years on a permanent basis (Ullah, Shamsuddoha, & Shahjahan, 2014: 201; Chakma, 2010).

The migration of Bengalis to the CHT accelerated after the division of Pakistan and the emergence of Bangladesh as a new state in 1971 following a violent independence struggle. The newly formed country ignored the demand for self-government by the tribal people and urged them to assimilate into mainstream Bengali society and culture (Chakma, 2010; Mohsin, 1997). It implemented a "rehabilitation" policy which involved settling war-ravaged refugees from the liberation war to the CHT (Chakma, 2010: 290). This rehabilitation process displaced some tribal people from their land (Chakma, 2010; Kamaluddin, 1980). Large-scale Bengali migration occurred after 1975 when successive governments encouraged landless Bengalis to migrate to the CHT (Ullah, Shamsuddoha, & Shahjahan, 2014: 203). The literature documents several phases of Bengali migration and settlement to the CHT:

> In phase one, from 1979–1981 about 30,000 families (around 100,000 people) came to the CHT and settled; in the second phase, another 100,000 people came immediately after the settlement; and in the third phase after 1982, another 250,000 people came and settled in the CHT.
>
> (Chakma, 2010; Ullah et al., 2014)

Some researchers argued that the successive governments of Bangladesh settled Bengali people in the CHT to relieve environmental displacement and demographic pressure in the rural areas, and to defuse the sentiment for self-determination of the tribal people in the CHT (Hafiz & Islam, 1993; Suhrke, 1997; Bächler, 1995; Reuveny, 2007).

Although around 400,000–500,000 Bengali people are estimated to have moved to the CHT through the settlement program, it is likely that others have migrated through social networks and family connections. The social network refers to marriage relations and knowing information about migration from someone who has already settled in the region (Hugo, 2013). The family connection is a source of migration in the CHT by which the settlement of one family member often encourages another member to come and settle permanently. This can be termed as the "fourth phase of migration," which is the result of the previous settlement and migration of the Bengali people. In this phase, government support is no longer provided or needed as people move to the CHT with the help of family members or social networks. These processes have resulted in the growth of the Bengali population in the CHT after independence, which has significantly changed the demographic composition in the region. The proportion of the tribal population has decreased as the Bengali population increased (Table 3.2).

The statistics in Table 3.2 reveal that the change in the ratio of the Bengali population to the tribal population in the CHT happened over a short period of time between 1979 and 1990 when the government pursued a settlement program. This population transformation is legal as the constitution of

Table 3.2 Bengali and tribal population ratio in the CHT, 1872–2011

Year	*Tribal*		*Bengali*		*Total Population*
	Population	*Percentage*	*Population*	*Percentage*	
1872	61,957	98	1097	2	63,054
1901	116,000	93	8762	7	124,762
1951	261,538	91	26,150	9	287,688
1961	339,757	88.28	45,322	11.77	385,679
1974	392,199	77.17	116,000	22.83	508,199
1981	441,776	59	304,873	41	746,649
1991	501,144	51	473,301	49	974,445
2001	592,977	44	740023	56	1,333,000
2011	845,541	53	752,690	47	1,598,231

Source: (Chakma, 2010: 291; Adnan, 2004: 15).

Bangladesh guarantees people's free movement and freedom to settle in any place (Mohsin, 1997: 113). Article 36 of the Bangladesh constitution states: "subject to any reasonable restrictions imposed by law in the public interest, every citizen shall have the right to move freely throughout Bangladesh, to reside and settle in any place therein and to leave and re-enter Bangladesh." However, the population movement to the CHT has been a source of significant conflict in the CHT. Tribal people were the majority until the 1980s, but now tribal and Bengali people are almost equal in number.

Conflict issue in Bangladesh: History, key issues and changes

Bangladesh has a "violent political culture," experiencing widespread political violence, particularly around elections – this feature has been attributed to its increasingly authoritarian form of democracy (Hassan & Nazneen, 2017). Political violence often involves the political youth and student cadres of the parties clashing in confrontations. The targets of political violence are prominently low and mid-ranking members of the political parties (Hassan & Nazneen, 2017).

Social, ethnic, and religious conflicts in the country have been seen to escalate when they were exploited with political intentions; however, they have been mostly latent thus far (BTI, 2018: 27). A new phase of Islamic radicalization and violence has risen during the past decade, with increased violent targeting of minorities, continued gender-based violence, and violence against labour rights protests and human rights activists (Herbert, 2019). Extremist groups pose prominent challenges to the state's monopoly on authorized force (BTI, 2018).

The most violent and conflict prone part of the country is the CHT. The genesis of the CHT conflict lies in the partition politics of the British government (Mohsin, 1997). The British colonial government, during its departure from the Indian subcontinent, annexed the CHT with Pakistan in a move that was opposed by the tribal people (Mohsin, 1995). Some tribal leaders hoisted the Indian flag in Rangamati during the partition. The Pakistani government responded not only by abolishing the special status of the CHT but also changing the social and economic environment by encouraging migration and construction projects such as Kaptai Dam (Arens et al., 2011; Mohsin, 1997; Shelly, 1992). The hope of the tribal people of gaining the right to self-determination remained unfulfilled even after the emergence of Bangladesh in 1971. This became evident when Manobendra Narayan Larma, the only elected member in the parliament from the CHT, tabled a list of demands during the formulation of the constitution of Bangladesh in 1972. The demands include autonomy for the CHT, including its own legislature; retention of the Regulation of 1900 Act in the constitution of Bangladesh; continuation of the "circle chief's office"[6]; and a constitutional provision restricting the amendment of Regulation of 1900 Act and imposing a ban on the influx of the non-hill people (Mohsin, 1997: 57–58). Ignoring the demands of the tribal leader, the new Bengali leadership urged the tribal people to merge into the Bengali culture and mainstream political system (Ahmed, 1993; Mohsin, 1997; Shelly, 1992). Since 1971 successive governments have undertaken significant economic, social, and infrastructure policies in the name of development; however, these policies have indirectly helped the successive governments of Bangladesh to consolidate their hegemony in the CHT (Ahmed, 1993; Mohsin, 1997; Nasreen & Togawa, 2002; Uddin, 2010).

The tribal leaders formed a regional platform the Parbatya Chattagram Jana Sanghati Samiti (PCJSS) to pursue their demands by political means, and when this was unsuccessful, a military wing Shanti Bahini (peace force) was formed to add pressure. The government interpreted the demand for self-determination as secessionist and identified the CHT as a national security problem. When the Shanti Bahini finally took up arms in 1976 (Mohsin, 1997), the Bangladesh state responded with wide-ranging militarization in the CHT. The entire region underwent a full-scale militarization, and a civil war between the Bangladesh army and the tribal people began (Mohsin, 1997; Chakma, 2010).

In the 1980s the Bangladesh state also settled landless, poor and environmentally affected people to the CHT, which is sometimes described as a secret device of the military government (Mohsin, 1997; Ullah, Shamsuddoha, Shahjahan, 2014: 203). This is a secret plan because the government of Bangladesh did not publish it officially but ordered the district commissioner of the Chittagong to implement the settlement policy in the CHT (Mohsin, 1997). This settlement policy has also been labelled a strategic hamlet (Chakma, 2010) and used by the military to resist the insurgency movement and to exterminate tribal people in the CHT (Chakma, 2010: 290). Bengali families were promised by the government 5 acres of hilly land, 4 acres of mixed land, and 2 and a half acres of paddy land (Chakma, 2010: 290). In some context, some Bengali settlers seized more land. Bengali families also received help and assistance from the Bangladesh army and, therefore, they sometimes worked with the army to resist the tribal separatist movement. Thus, the Bengali migrants on many occasions have been part of the CHT conflict.

Moreover, allocation of land to the Bengali people by the civil administration generated resentment, anger, and anguish among the tribal people. Adnan (2007) argued that Bengali migrants have been the "soft target" of the Shanti Bahini, and on many occasions, the Shanti Bahini attacked the Bengali settlers to force them to leave the CHT. As a result, both parties in the conflict targeted civilians, property, and objects (Adnan, 2007: 15). The army used the Bengali settlers as a human shield, and the tribal people targeted settlers to make them leave their land. This conflict situation also forced some tribal families to move deeper into the jungle to avoid conflict and violence. According to Mohsin (2000, 115), "Bengali settlement program has deepened the alienation of the Hill people as many tribal people have lost their land and control over the forest resources." The conflict, capturing of land, communal violence, and human rights violations continued until the CHT Peace Accord was enacted between the government of Bangladesh and the PCJSS in 1997. This accord was the result of a series of discussions, negotiations, and multilevel consultations between the government led by prime minister Sheikh Hasina and tribal leaders (Mohsin, 2003: 13).

The CHT Accord ended the armed conflict between the Bangladesh army and the Shanti Bahini and raised hopes that people would soon enjoy freedom and rights in their land. The members of Shanti Bahini surrendered their arms to the government of Bangladesh in return for promised jobs in the security forces and financial benefits. The Peace Accord as a whole included some principles, such as (1) land rights, (2) revival of cultural identity, (3) rehabilitation of refugees, and (4) withdrawal of the temporary military camp, to restore the authority of the tribal people over the land and resources as well as enjoyment of human rights in the CHT (Jamil & Panday, 2008: 471). The issue of internally displaced people (IDP) remained unresolved as the tribal leaders demanded the return of Bengali settlers who had migrated in the 1980s. The tribal leaders also demanded the repatriation of all tribal families displaced during the war. However, these demands of the tribal leaders have not been enacted, and instead, residency certificates were provided to all Bengali settlers by the Deputy Commission (DC) of the CHT (Mohsin, 2003: 73). This decision frustrated the tribal leaders who on occasions boycotted meetings with the government authorities working in the CHT (Jamil & Panday, 2008; Mohsin, 2003). The post Peace Accord situation of the CHT is characterized by conflict and violence in what has been described as a violent peace (Chakma & d'Costa, 2013), and peace without justice (Mohsin, 2003: 54). The main reasons for the continuation of conflict are resource scarcity, land grabbing and competition between the communities for social position (Ullah, Shamsuddoha, & Shahjahan, 2014; Adnan, 2004; Panday & Jamil, 2009). Moreover, the military has remained entrenched in the CHT, and military and civil bureaucrats capture resources and exacerbate the competition for land and forest, fuelling further related conflicts.

Some researchers have drawn connections between environment and climate change–induced migration and violent conflict in the region. Hafiz and Islam (1993) first noted that people were driven to the CHT by environmental problems and influenced the conflict (Hafiz and Islam 1993). Other papers followed, citing the connection between the environmentally induced migration and conflict in the CHT (Suhrke, 1997; Lee, 1997; Reuveny, 2007; Baechler, 1998). They argue that the conflict escalated because Bengali people put pressure on the land and destabilised the way of life of the tribal people, raising fears of extinction. From this perspective, the CHT conflict is described as a demographic and environmentally induced conflict (Bächler 1995: 25). Smith and Vivekananda (2007: 16) have argued that climate change effects have an indirect role in escalating the conflict in the CHT. The climate change–induced Bengali migrants complicated the CHT conflict when they migrated and settled in the region. Thus, over time, Bengali migration and settlement in the CHT have pitched settlers and tribal people against each other as the primary agents in the conflict in the post Peace Accord era (Mohsin, 2003). In addition to conflict over land, religious differences have emerged as a source of conflict, as Muslim people build more mosques and madrasas for their prayer and Islamic education. The aggressive pursuit of Islamic religious interests has aggravated the tribal people, who are predominantly Buddhist (Joarder & Miller, 2013). For example, on many occasions unidentified people vandalized the temples of the Buddhist people in the CHT, which caused communal conflict between Bengali and tribal people (Kapaeeng Foundation, 2017).

The rise of extremist political groups in the CHT has also contributed to the current conflict. For example, the United People's Democratic Front (UPDF) has formed from the *Jana Sambati Samiti* (JSS) who oppose the Peace Accord, Bengali settlement, and migration in the CHT. The UPDF is seeking secession of the CHT from Bangladesh (Mohsin, 2003; Uddin, 2011). Its antisettlement stance has fueled the conflict between the UPDF and settlers, as well as with the Bangladesh army. In response, the Bengali settlers have organised and formed a political platform named *Somo Adikhar Andolon* (SAA: Equal Rights Movement) to fight for their survival and for equal rights in all spheres of life in the CHT. The SAA was formed in 2001 by the settlers who believe the Peace Accord has made them second-class citizens in the CHT. The SAA movement has opposed the PCJSS for signing the Peace Accord and has carried out several agitations demanding the annulment of the 1997 accord (Panday & Jamil, 2009; Mohsin, 2003). Thus, the emergence of extremist groups since the Peace Accord has led to new confrontations. There are many reasons for this, but key to understanding these divisions among the groups and community in the CHT is the lack of civil society involvement in the peace negotiations (Chakma & D'Costa, 2013) and lack of implementation of the accord. A study has found that implementation of the principles of the Peace Accord by the signatory parties helps to reduce the incidents of violence and conflict in the post-conflict society (Joshi & Quinn, 2016).

In summary, the CHT conflict is one of intractable ethnopolitical conflicts which has a colonial legacy, as well as a long history of the denial of identity of the tribal people. The hegemony of the ruling classes has been established over the tribal people from the British to the current Bangladesh period. While the construction of the hegemony and denial of identity constituted the key issues of the CHT conflict, the issues of conflict have been transformed and diversified over time. Now the Bengali and tribal polarization is an established phenomenon with cleavages in culture, religion, social, and political aspirations between the two groups of people. Mistrust and lack of proper reconciliation processes have aggravated the conflict. This situation has created fertile ground to form extremist groups who are intimidating and killing people and violating their human rights. The unfulfilled promises of the government of Bangladesh and the growing rise of group consolidation among the Bengali settlers are also hindering the current peacebuilding processes in the region.

Conclusion

There is little doubt that climatic events accompanied by socioeconomic and political factors are causing livelihood failure and human displacement in the low-lying and rural areas in Bangladesh. Displaced people migrated to the cities and also to the CHT during the 1980s with the direct and indirect assistance of the government. Successive governments targeted these environmentally induced migrants during the time of Bengali settlement to the CHT in order to defuse the conflict and reduce the vulnerabilities in the climate affected areas. The Bengali settlers have been given land by the government for housing and agriculture (Mohsin, 1997; Chakma, 2010), putting them in competition with the tribal people in the region. Resource competition accompanied by social and political variables, such as increasing growing population, poverty, ethnic divisions, past conflict and deprivation combine to generate violence and conflict between the original and migratory people in the CHT.

Even though many studies have cited the CHT case as a conflict situation due to the environmental migration, no study has yet explored the relationship of climatic change, migration, and conflict in the CHT in Bangladesh in any depth. In this story of the CHT, the key elements of existing theories of environmental and climate conflict are all present: climate change events, forced migration, resource scarcity (Homer-Dixon, 1999), population pressure (Kahl, 2006; Goldstone, 2002), ethnic division, and marginalization of minority groups (Gurr, 2000; Cederman, Wimmer, & Min, 2010). In the following chapters, the relationship between climate change, migration, and conflict will be explored in detail based on the analytical framework presented in Figure 2.3. It is important to note that this causal relationship is not inevitable in all contexts where there are climate changes effects, but it is applicable in the case of Bangladesh to explain the migration of the Bengali settlers and CHT conflict.

Notes

1 A decimal (also spelled decimel) is a unit of area in India and Bangladesh approximately equal to 1/100 acre (40.46 m^2). 1 decimal is equal to 435.6 square feet.
2 Sidr was a devastating cyclone which struck the southwest coast of Bangladesh on 15 November 2007 with wind speeds of 240 kilometers per hours. This cyclone resulted in tidal waves up to 5 meters high and surges of up to 6 meters. This was one of the most devastating cyclones in Bangladesh, which damaged 2.3 million households, and 1 million people were seriously affected. It was estimated that cyclone Sidr cost nearly US$ 1.1 billion (Government of Bangladesh, 2008; Paul, 2010).
3 Cyclone Nargis hit Myanmar but flowed over some parts of Bangladesh in 2008. This cyclone caused massive displacement and property destruction in Bangladesh (Mallick & Vogt, 2012).
4 In the coastal area of Bangladesh, land is being destroyed by flood water or river-bank erosion every year. However, new land emerges in coastal areas, which is known as 'char'. Around 52 square kilometers of new char land is formed every year in Bangladesh (Shamsuddoha, Shahid Ullah, Shahjahan, 2014).
5 Pakistan after its emergence as an independent state introduced Islamic law, declaring the state as an Islamic Republic of Pakistan in 1956. This declaration introduced the policy of expansion of Muslim culture across the country. The CHT, as a non-Muslim region, has been the victim of this Islamic state.
6 The circle chief is the traditional administrative system in the CHT among the tribal people, who enjoys power in appointing headmen, collecting taxes, and helping government for the use of land. Although this is the top administrative system among the tribal people, it currently works under the District Commissioners.

References

Adnan, S. (2004). *Migration land alienation and ethnic conflict: Causes of poverty in the Chittagong Hill Tracts of Bangladesh*. Dhaka: Research & Advisory Services.
Adnan, S. (2007). Migration, discrimination and land alienation: Social and historical perspectives on the ethnic conflict in the Chittagong Hill Tracts of Bangladesh. *Contemporary Perspectives*, *1*(2), 1–28.
Afsar, R. (2003). *Internal migration and the development nexus: The case of Bangladesh*. Paper presented at the *Regional Conference on Migration, Development and Pro-Poor Policy Choices in Asia*, 22–24 June 2003 in Dhaka, Bangladesh. https://www.researchgate.net/publication/228916027_Internal_Migration_and_the_Devopment_Nexus_The_Case_of_Bangladesh
Agrawala, S., Ota, T., Ahmed, A. U., Smith, J., & Van Aalst, M. (2003). *Development and climate change in Bangladesh: Focus on coastal flooding and the Sundarbans* (pp. 1–49). Paris: Organisation for Economic Co-operation and Development (OECD). http://www.oecd.org/dataoecd/46/55/21055658.pdf

Ahmad, S. (2015). Housing demand and housing policy in urban Bangladesh. *Urban Studies*, *52*(4), 738–755.

Ahmed, A. (1993). Ethnicity and insurgency in the Chittagong Hill Tracts region: A study of the crisis of political integration in Bangladesh. *Journal of Commonwealth & Comparative Politics*, *31*(3), 32–66.

Ahmed, A. U. (2006). *Bangladesh climate change impacts and vulnerability: A synthesis*. Bangladesh: Climate Change Cell, Department of Environment https://www.preventionweb.net/files/574_10370.pdf

Ahmed, S., & Meenar, M. (2018). Just sustainability in the global South: A case study of the megacity of Dhaka. *Journal of Developing Societies*, 34(4), 401–424.

Ahsan, R. (2019). Climate-induced migration: Impacts on social structures and justice in Bangladesh. *South Asia Research*, *39*(2), 184–201.

Ahsan, R., Kellett, J., & Karuppannan, S. (2014). Climate induced migration: Lessons from Bangladesh. *The International Journal of Climate Change: Impacts and Responses*, *5*, 1–14.

Ahsan, R., Kellett, J., & Karuppannan, S. (2016). Climate migration and urban changes in Bangladesh. In R. Shaw, A. Rahman, A. Surjan, & G. Parvin (Eds.), *Urban disasters and resilience in Asia* (pp. 293–316). Amsterdam: Butterworth-Heinemann.

Amin, M. A. (2018, November 23). Dhaka, Chittagong Destination of 80% internal migrants. https://www.dhakatribune.com/opinion/special/2018/11/24/dhaka-chittagong-destination-of-80-internal-migrants?__cf_chl_managed_tk__=pmd_v_u73US.A4irZECh9OCQDaxEU01gnqnWR_yfflS8I1Y-1633510326-0-gqNtZGzNAyWjcnBszQcR

Amjad, K. (2021). Factors of climate migration of poor slum dwellers in Dhaka city. *International Journal of Earth Sciences Knowledge and Applications*, *3*(3), 273–288.

Arens, C., Wang-Helmreich, H., Hodes, G. S., & Burian, M. (2011). *Assessing support activities by international donors for CDM development in Sub-Saharan Africa with focus on selected least developed countries*. Wuppertal/Hamburg: Wuppertal Institute and GFA Envest. http://www.jikobmub.de/files/basisinformationen/publikationen/application/pdf/donor_activities_barrierefrei.pdf

Arsenault, M. P., Azam, M. N., & Ahmad, S. (2015). Riverbank erosion and migration in Bangladesh's char lands. In B. Mallick & B. Etzold (Eds.), *Environment, migration and adaptation: Evidence and politics of climate change in Bangladesh* (pp. 41–62). Dhaka: AHDPH.

Bächler, G. (1995). The anthropogenic transformation of the environment: A source of war? Historical background, typology and conclusions. In K. P. Spillmann, & G. Bächler (Eds.), *Environment and conflicts project (ENCOP)*, (pp. 11–27). Zurich-Switzerland: Center for Security Studies and Conflict Research, Swiss Federal Institute of technology.

Baechler, G. (1998). Why environmental transformation causes violence: A synthesis. *Environmental Change and Security Project Report*, *4*(1), 24–44.

Bangladesh Metrological Department. (2017). *Weather report*. http://bmd.gov.bd/p/Reports/

Bertelsmann Stiftung (BTI) (2018) *Country Report — Bangladesh*. Gütersloh: Bertelsmann Stiftung. https://www.btiproject.org/fileadmin/files/BTI/Downloads/Reports/2018/pdf/BTI_2018_Bangladesh.pdf

Black, R., Arnell, N. W., Adger, W. N., Thomas, D., & Geddes, A. (2013a). Migration, immobility and displacement outcomes following extreme events. *Environmental Science & Policy*, 27(1), S32–S43. https://doi.org/10.1016/j.envsci.2012.09.001

Black, R., Kniveton, D., & Schmidt-Verkerk, K. (2013b). Migration and climate change: Toward an integrated assessment of sensitivity. In T. Faist & J. Schade (Eds.), *Disentangling migration and climate change: Methodologies, political discourses and human rights* (pp. 29–53). Dordrecht: Springer Netherlands. Imprint: Springer. Black.

Black, R., Kniveton, D., Skeldon, R., Coppard, D., Murata, A., & Schmidt-Verkerk, K. (2008). *Demographics and climate change: Future trends and their policy implications for migration*. Working Paper, Development Research Centre on Migration, Globalization and Poverty. Brighton: University of Sussex. http://www.migrationdrc.org/publications/working_papers/WP-T27.pdf

Bose, S. (2013). Sea-level rise and population displacement in Bangladesh: Impact on India. *Maritime Affairs: Journal of the National Maritime Foundation of India*, 9(2), 62–81. https://doi.org/10.1080/09733159.2013.848616

Brammer, H. (2014). Bangladesh's dynamic coastal regions and sea-level rise. *Climate Risk Management*, *1*, 51–62.

Brouwer, R., Akter, S., Brander, L., & Haque, E. (2007). Socioeconomic vulnerability and adaptation to environmental risk: A case study of climate change and flooding in Bangladesh. *Risk Analysis*, *27*(2), 313–326. https://doi.org/10.1111/j.1539-6924.2007.00884.x

Bryan, G., Chowdhury, S., & Mobarak, A. M. (2014). Underinvestment in a profitable technology: The case of seasonal migration in Bangladesh. *Econometrica*, *82*(5), 1671–1748. https://doi.org/10.3982/ECTA10489

BTI. (2018). *BTI 2018 Country Report: Bangladesh*. Gütersloh: Bertelsmann Stiftung.

Call, M. A., Gray, C., Yunus, M., & Emch, M. (2017). Disruption, not displacement: Environmental variability and temporary migration in Bangladesh. *Global Environmental Change*, *46*, 157–165. https://dx.doi.org/10.1016%2Fj.gloenvcha.2017.08.008

Castellano, R., Dolšak, N., & Prakash, A. (2021). Willingness to help climate migrants: A survey experiment in the Korail slum of Dhaka, Bangladesh. *PloS One*, *16*(4), e0249315.

Cederman, L. E., Wimmer, A., & Min, B. (2010). Why do ethnic groups rebel?. New data and analysis *World Politics*, *62*(1), 87–119.

Chakma, B. (2010). The post-colonial state and minorities: Ethnocide in the Chittagong Hill Tracts, Bangladesh. *Commonwealth & Comparative Politics*, *48*(3), 281–300. http://www.tandfonline.com/action/showCitFormats?doi=10.1080/14662043.2010.489746

Chakma, K., & D'Costa, B. (2013). The Chittagong Hill Tracts: Diminishing violence or violent peace, In E. Aspinall, R. Jeffrey, & A. Regan, (Eds.), *Diminishing conflicts in Asia and the Pacific: Why some subside and others don't* (pp. 137–149). Abingdon, Oxon: Routledge.

Chatterjee, R. (1987). Cotton handloom manufactures of Bengal, 1870-1921. *Economic and Political Weekly*, *22*(25), 988–997.

Chowdhury, M. B. (2011). Remittances flow and financial development in Bangladesh. *Economic Modelling*, *28*(6), 2600–2608.

Comprehensive Disaster Management Programme (CDMP II). (2013). *Vulnerability to climate induced droughts: Scenario and impacts*. Dhaka: Ministry of Disaster Management and Relief. http://www.bd.undp.org/content/bangladesh/en/home/library/crisis_prevention_and_recovery/vulnerability-to-climate-induced-drought--scenario---impacts.html

Davis, K. F., Bhattachan, A., D'Odorico, P., & Suweis, S. (2018). A universal model for predicting human migration under climate change: Examining future sea level rise in Bangladesh. *Environmental Research Letters*, *13*(6), 064030.

Etzold, B., Ahmed, A. U., Hassan, S. R., & Neelormi, S. (2014). Clouds gather in the sky, but no rain falls. Vulnerability to rainfall variability and food insecurity in northern Bangladesh and its effects on migration. *Climate and Development*, *6*(1), 18–27. https://doi.org/10.1080/17565529.2013.833078

Findlay, A., & Geddes, A. (2011). Critical views on the relationship between climate change and migration: Some insights from the experience of Bangladesh. In E. Piguet, A. Pécoud, & P. de Guchteneire (Eds.), *Migration and climate change* (pp. 138–159). Paris: UNESCO Publishing.

Gray, C. L., & Mueller, V. (2012). Natural disasters and population mobility in Bangladesh. *Proceedings of the National Academy of Sciences*, *109*(16), 6000–6005.

Gurr, T. R. (2000). Ethnic warfare on the wane. *Foreign Affairs*, *79*(3), 52–64.

Hafiz, M. A., & Islam, N. (1993). *Environmental degradation and intra/interstate conflicts in Bangladesh*. ETH Zurich: Center for Security Studies (CSS).

Hassan, M., & Nazneen, S. (2017). Violence and the breakdown of the political settlement: An uncertain future for Bangladesh?. *Conflict, Security & Development*, *17*(3), 205–223.

Hassani-Mahmooei, B., & Parris, B. W. (2012). Climate change and internal migration patterns in Bangladesh: An agent-based model. *Environment and Development Economics*, *17*(6), 763–780. https://doi.org/10.1017/S1355770X12000290

Herbert, S. (2019). *Conflict analysis of Bangladesh*. K4D Helpdesk Report 599. Brighton, UK: Institute of Development Studies.

Homer-Dixon, T. F. (1999). *Environment, scarcity, and violence*. Princeton, NJ: Princeton University Press.

Hugo, G. (2013). *Migration and climate change*. Cheltenham, UK: Edward Elgar Publishing.

Intergovernmental Panel on Climate Change (IPCC). (2014). *Climate change 2014: Synthesis report*. Geneva 2, Switzerland Intergovernmental Panel on Climate Change. http://ipcc.ch/pdf/assessmentreport/ar5/syr/AR5_SYR_FINAL_All_Topics.pdf

Islam, N., Khan, A. U., & Khan, A. A. M. (1975). Survey of Urban Squatters in Dacca, Chittagong and Khulna, 1974: Draft Report Prepared for the Urban Development Directorate, Government of the People's Republic of Bangladesh; Report. Centre for Urban Studies, Department of Geography, University of Dacca.

Jamil, I., & Panday, P. K. (2008). The elusive peace accord in the Chittagong Hill Tracts of Bangladesh and the plight of the indigenous people. *Commonwealth & Comparative Politics*, *46*(4), 464–489. https://doi.org/10.1080/14662040802461141

Joarder, M. A. M., & Miller, P. W. (2013). Factors affecting whether environmental migration is temporary or permanent: Evidence from Bangladesh. *Global Environmental Change*, *23*(6), 1511–1524. https://doi.org/10.1016/j.gloenvcha.2013.07.026

Joshi, M., & Quinn, J. M. (2016). Watch and learn: Spill-over effects of peace accord implementation on non-signatory armed groups. *Research & Politics*, *3*(1), 1–7. https://doi.org/10.1177%2F2053168016640558

Kahl, C. H. (2006). *States, scarcity, and civil strife in the developing world*. Princeton, Woodstock: Princeton University Press.

Kamaluddin, S. (1980). A tangled web of insurgency. *Far Eastern Economic Review*, *23*(5), 80.

Kapaeeng Foundation (2017). Intervention of Bablu Chakma on agenda item 4 of 16th session of the UNPFii. http://www.kapaeeng.org/intervention-of-bablu-chakma-on-agenda-item-4-of-16th-session-of-the-unpfii/

Karim, M. F., & Mimura, N. (2008). Impacts of climate change and sea-level rise on cyclonic storm surge floods in Bangladesh. *Global Environmental Change*, *18*(3), 490–500.

Kartiki, K. (2011). Climate change and migration: A case study from rural Bangladesh. *Gender & Development*, *19*(1), 23–38. http://dx.doi.org/10.1080/13552074.2011.554017

Kemper, R. V. (1989). Urbanization in Bangladesh: Historical development and contemporary crisis. *Urban Anthropology and Studies of Cultural Systems and World Economic Development*, *18*, 365–392.

Khan, A. A. M. (1982). Rural-urban migration and urbanization in Bangladesh. *Geographical Review*, *72*, 379–394.

Khan, H. (2008). Challenges for sustainable development: Rapid urbanization, poverty and capabilities in Bangladesh.

Khan, S. I., Sarker, M. N. I., Huda, N., Nurullah, A. B. M., & Zaman, M. R. (2018). Assessment of new urban poverty of vulnerable urban dwellers in the context of sub-urbanization in Bangladesh. *The Journal of Social Sciences Research*, *4*(10), 184–193.

Kharat, R. S. (2003). *From internal displacement to refugees: The trauma of Chakmas in Bangladesh*. Paper presented at the Research Paper is being presented at *Researching International Displacement: State of the Art International Conference on IDPs*, 7–8 February 2003, Trondheim, Norway.

Lee, S.-W. (1997). Not a one-time event: Environmental change, ethnic rivalry, and violent conflict in the Third World. *The Journal of Environment & Development*, *6*(4), 365–396.

Lee, S.-W. (2001). *Environment matters: Conflicts, refugees & international relations*. Seoul and Tokyo: World Human Development Institute Press.

Lein, H. (2000). Hazards and “forced” migration in Bangladesh. *Norsk Geografisk Tidsskrift*, *54*(3), 122–127. https://doi.org/10.1080/002919500423735

Levene, M. (1999). The Chittagong Hill Tracts: A case study in the political economy of creeping genocide. *Third World Quarterly*, *20*(2), 339–369.

Mallick, B., & Vogt, J. (2012). Cyclone, coastal society and migration: Empirical evidence from Bangladesh. *International Development Planning Review*, *34*(3), 217–240. https://doi.org/10.3828/idpr.2012.16

Mannan, M. A. (2018). Burden of disease on the urban poor: A study of morbidity and utilisation of healthcare among slum dwellers in Dhaka City.

Maplecroft, V. (2011). *Climate change vulnerability index 2016*. Bonn: Climate Change and Environmental Risk Atlas.

Mazumder, M. S. U., & Wencong, L. (2012). Monga vulnerability in the northern part of Bangladesh. *African Journal of Agricultural Research*, *7*(3), 358–366.

Mehedi, H., Nag, A. K., & Farhana, S. (2010). Climate induced displacement: Case study of cyclone Aila in the southwest coastal region of Bangladesh. *Humanitywatch*. Khulna, 1–28.

Minar, M. H., Hossain, M. B., & Shamsuddin, M. D. (2013). Climate change and coastal zone of Bangladesh: Vulnerability, resilience and adaptability. *Middle-East Journal of Scientific Research*, *13*(1), 114–120.

Mirza, M. M. Q., & Ahmad, Q. K. (Eds.). (2005). Climate change and water resources in South Asia: An Introduction. In M. M. Mirza, & Q. K. Ahmed (Eds.), *Climate change and water resources in South Asia*, (pp. 1–21). Leiden, Natherlands: A.A. Balkema Pub.

Mirza, M. M. Q., Warrick, R. A., & Ericksen, N. J. (2003). The implications of climate change on floods of the Ganges, Brahmaputra and Meghna rivers in Bangladesh. *Climatic Change*, *57*(3), 287–318.

Mohsin, A. (1995) *The politics of nationalism: The case of the Chittagong Hill Tracts, Bangladesh*. Doctoral dissertation, University of Cambridge.

Mohsin, A. (1997). Dhaka: University Press Limited (UPL).

Mohsin, A. (2000). Identity, politics and hegemony. *Identity, Culture and Politics*, *1*(1). http://calternatives.org/resource/pdf/IDENTITY,%20POLITICS%20AND%20HEGEMONY-The%20Chittagong%20Hill%20Tracts,%20Bangladesh.pdf

Mohsin, A. (2003). *The Chittagong Hill Tracts, Bangladesh: On the difficult road to peace*. Boulder, CO: L. Rienner.

Mutton, D., & Haque, C. E. (2004). Human vulnerability, dislocation and resettlement: Adaptation processes of river-bank erosion-induced displacees in Bangladesh. *Disasters*, *28*(1), 41–62.

Myers, N., & Kent, J. (1995). *Environmental exodus: An emergent crisis in the global arena*. Washington, DC: Climate Institute. http://climate.org/archive/PDF/Environmental%20Exodus.pdf

Nasreen, Z., & Togawa, M. (2002). Politics of development: 'Pahari-Bengali' discourse in the Chittagong Hill Tracts. *Journal of International Development and Cooperation*, *9*(1), 97–112.

Neumann, B., Vafeidis, A. T., Zimmermann, J., & Nicholls, R. J. (2015). Future coastal population growth and exposure to sea-level rise and coastal flooding-a global assessment. *PloS One*, *10*(3), e0118571.

Panday, P. K., & Jamil, I. (2009). Conflict in the Chittagong Hill Tracts of Bangladesh: An unimplemented accord and continued violence. *Asian Survey*, *49*(6), 1052–1070.

Paul, B. K. (2005). Evidence against disaster-induced migration: The 2004 tornado in north-central Bangladesh. *Disasters*, *29*(4), 370–385.

Paul, S. K., & Routray, J. K. (2010). Flood proneness and coping strategies: The experiences of two villages in Bangladesh. *Disasters*, *34*(2), 489–508.

Pender, J. S. (2008). *What is climate change? and how it will effect Bangladesh*. Briefing paper (final draft). Dhaka, Bangladesh: Church of Bangladesh Social Development Programme.

Penning-Rowsell, E. C., Sultana, P., & Thompson, P. M. (2013). The "last resort"? Population movement in response to climate-related hazards in Bangladesh. *Environmental Science & Policy*, *27*, S44–S59. https://doi.org/10.1016/j.envsci.2012.03.009

Quader, M., Khan, A., & Kervyn, M. (2017). Assessing risks from cyclones for human lives and livelihoods in the coastal region of Bangladesh. *International Journal of Environmental Research and Public Health*, *14*(8), 831.

Quasem, M. A. (2011). Conversion of agricultural land to non-agricultural uses in Bangladesh: Extent and determinants. *The Bangladesh Development Studies*, *34*(1), 59–85.

Quencez, M. (2011). Floods in Bangladesh and migration to India. In F. Gemenne, P. Brücker, & D. Ionesco (Eds.), *The state of environmental migration* (pp. 57–71). https://publications.iom.int/system/files/pdf/state_environmental_migration_2011.pdf

Rahman, S., & Rahman, M. (2009). Impact of land fragmentation and resource ownership on productivity and efficiency: The case of rice producers in Bangladesh. *Land Use Policy*, *26*(1), 95–103.

Raleigh, C., & Jordan, L. (2010). Climate change and migration: Emerging patterns in the developing world. In R. Mearns & A. Norton (Eds.), *The social dimensions of climate change: Equity and vulnerability in a warming world* (pp. 103–133). Washington, DC: World Bank.

Rayhan, I., & Grote, U. (2007). Coping with floods: Does rural-urban migration play any role for survival in rural Bangladesh. *Journal of Identity and Migration Studies*, *1*(2), 82–98.

Reuveny, R. (2007). Climate change-induced migration and violent conflict. *Political Geography*, *26*(6), 656–673. https://doi.org/10.1016/j.polgeo.2007.05.001.

Rigaud, K. K., de Sherbinin, A., Jones, B., Bergmann, J., Clement, V., Ober, K., … Heuser, S. (2018). *Groundswell: Preparing for internal climate migration*. Washington, DC: The World Bank.

Shamsuddoha, M., & Chowdhury, R. K. (2007). *Climate change impact and disaster vulnerabilities in the coastal areas of Bangladesh*. COAST Trust, Dhaka.

Shelly, M. R. (1992). *The Chittagong Hill Tracts of Bangladesh: The untold story*. Dhaka, Bangladesh: Centre for Development Research, Bangladesh.

Siddiqui, T. (2005). International labour migration from Bangladesh: A decent work perspective. *Policy Integration Department Working Paper, 66*. Geneva: International Labour Office. http://citeseerx.ist.psu.edu/viewdoc/download?doi=10.1.1.471.6259&rep=rep1&type=pdf

Siddiqui, T., & Billah, M. (2014). Adaptation to climate change in Bangladesh: Migration the missing link. In S. Vachani, & J. Usmani (Eds.), *Adaptation to climate change in Asia* (pp. 117–141). Cheltenham: Edward Elgar. https://doi.org/10.4337/9781781954737

Smith, D., & Vivekananda, J. (2007). *A climate of conflict: The links between climate change, peace and war*. London: International Alert. https://www.preventionweb.net/publications/view/7948

Streatfield, P. K., & Karar, Z. A. (2008). Population challenges for Bangladesh in the coming decades. *Journal of Health, Population, and Nutrition*, 26(3), 261.

Suhrke, A. (1997). Environmental degradation, migration, and the potential for violent conflict. In N.P. Gleditsch (Eds.), *Conflict and the environment* (pp. 255–272). NATO ASI Series (Series 2: Environment), vol 33. Dordrecht: Springer.

Tripura, S. B. (2008). *Blaming jhum, denying jhumia: Challenges of indigenous peoples land rights in the Chittagong Hill Tracts (CHT) of Bangladesh: A case study on Chakma and Tripura*. Masters thesis, Universitetet i Tromsø, Norway. https://munin.uit.no/handle/10037/1535

Uddin, N. (2010). Politics of cultural difference: Identity and marginality in the Chittagong hill tracts of Bangladesh. *South Asian Survey*, *17*(2), 283–294.

Uddin, N. (2011). Decolonising ethnography in the field: An anthropological account. *International Journal of Social Research Methodology*, *14*(6), 455–467.

Ullah, M. S., Shamsuddoha, M., & Shahjahan, M. (2014). The viability of the Chittagong Hill Tracts as a destination for climate-displaced communities in Bangladesh. In S. Leckie (Ed.), *Land solutions for climate displacement* (pp. 215–247). London: Routledge.

University Corporation of Atmospheric Research (UCAR). (2018). *Sea-level change in Bangladesh*. https://scied.ucar.edu/sea-level-change-bangladesh

Van Schendel, W., Mey, W., & Dewan, A. K. (2000). *The Chittagong Hill Tracts: Living in a borderland*. Bangkok: White Lotus.
Walsham, M. (2010). *Assessing the evidence: Environment, climate change and migration in Bangladesh*. Geneva: International Organization for Migration (IOM), Regional Office for South Asia. https://www.iom.int/jahia/webdav/site/myjahiasite/shared/shared/mainsite/events/docs/Assessing_the_Evidence_Bangaldesh.pdf
World Population Review. (2018). *Bangladesh population*. http://worldpopulationreview.com/countries/bangladesh-population/
Zaman, M. (1989). The social and political context of adjustment to riverbank erosion hazard and population resettlement in Bangladesh. *Human Organization*, *48*(3), 196–205.

4 The climate change and migration interplay in Bangladesh

Introduction

> Climate [change] influences human migration. Sometimes the influence is direct and obvious ... Sometimes its influence is more subtle ... Sometimes climate's influence is nested so deeply in interwoven chains of past events that we no longer notice how it has shaped where, how, and why we live in the places we do.
>
> (McLeman, 2014: 1)

> Migration options, and migrant agency are shaped and constrained by culture, economic, political, and social factors that operate at multiple scales, often well beyond the direct control or influences of the individuals, household, or local population.
>
> (McLeman, 2014: 230)

Migration – both internal and international – is a continuous process in Bangladesh. People migrate from one place to another to avoid vulnerabilities, shocks, and uncertainties. Sometimes, people migrate to the urban areas for a better life and livelihood. This chapter analyses the mechanisms, causes, and interplay of the migration in the cities and in the CHT in Bangladesh, and climatic events contribute to the migration and settlement. In the case of the CHT, this migration and settlement is viewed as political migration by many researchers, but some scholars have argued that it has been influenced by environment and climate change factors and poverty and that it has complicated the Chittagong Hill Tracts (CHT) conflict (Hafiz & Islam, 1993; Lee, 2001; Reuveny, 2007). This chapter offers empirical evidence demonstrating the relevance of climatic events in the migration of Bengali people to the CHT as well as how poor and climate affected people migrate to the urban areas. It draws on interview data to explore the impacts of climate change on migration decisions and processes. The key informants have also shared their ideas about the connection of climate change and migration processes in Bangladesh. The argument this chapter presents is that climate change effects

DOI: 10.4324/9781003430629-4

and poverty constituted the push factors for the migration of the people. The work opportunity in the urban areas and setting up almost all industries in the urban centers encourages the climate affected people to come and find scope of work in the city centers. The household work opportunities in the cities also encourage the people to come and settle in the slums so that they could at least work as household workers.

In the case of migration and settlement of the Bengali people to the CHT, the proposed opportunities by the government (e.g., the possibility of getting land) inspired most of the Bengali respondents to migrate to this region, even though it was in many ways an adverse destination due to conflict, the unfamiliar hilly environment, distance and remoteness, and lack of facilities, including road, medical, health, and education infrastructure.

Climate change and migration interplay

Push and pull factors effects

In Bangladesh, internal migration has happened due to push and pull factors. Climate changes and natural disasters displace people from their land. On the other hand, people also migrate for better opportunities. A study conducted by Islam, Schech, and Saikia (2020: 5) find the following main reasons (Table 4.1) for their decision to migrate from their place of origin to the CHT in Bangladesh.

The figure shows that migration from the other parts of Bangladesh to the CHT has occurred both for the push factors and pull factors. The climate change events forced 51 percent of the respondents to migrate to the CHT. On the other hand, the opportunities for getting land and resources for leading their life also motivated people to migrate to the CHT. A majority of the respondents outline that poverty and livelihood failure constituted the core causes of migration from their place of origin.

Different studies also show the relevance of this information on the impact of climatic events in the migration decision of the people. For example, the Environmental Justice Foundation (2018) shows that up to 50 percent of

Table 4.1 Reasons for migrating of the Bengali people to the CHT

Reasons for migration	*Percentage*
Poverty	69.3%
More income	45%
Political problem	3.3%
Lack of housing	45.3%
Climate change events	51%
Lack of livelihood	79.3%
Debt	6.7%

Source: Islam, Schech, & Saikia, 2020: 5.

slum dwellers were forced to leave their homes due to climatic events. There migrants are also called economic migrants because they settle in the slums mainly for economic reasons. Another study found an increasing number of climate migrants (Figure 4.1) in the urban areas in Bangladesh.

Figure 4.1 shows that the number of climate migrants to the urban areas is rapidly increasing in Bangladesh. The most frequently cited reasons for migrating were related to livelihood, income, and climatic events. Climate affected people become unable to make a living in their place of origin and migrate in the hope of improving their income and taking shelter.

Indeed, the interplay between climate change and migration is a complex process in Bangladesh. As discussed in Chapter 3, climatic events in Bangladesh have adverse effects on livelihood, housing, and other basic needs and cause displacement and migration (Penning-Rowsell et al., 2013; Mallick & Vogt, 2012; Joarder & Miller, 2013). As stated in Chapter 2, multiple factors such as climate events, poverty, expectation of more income, and avoiding resultant vulnerabilities play a role in migration decisions (Black, Adger, et al., 2011; McLeman, 2014). In the case of internal migration in Bangladesh, multiple causes such as poverty, frequent climate events, debt, and crop failure determine the migration decisions of the people (Joarder & Miller, 2013; Siddiqui & Billah, 2014). Chapter 3 showed that Bangladesh, as the most climate affected country, is facing annual floods, frequent cyclones, and storm surges. In the last two decades, the country has experienced five devastating cyclones and six floods which impacted the economy and human casualties (Penning-Rowsell et al., 2013: 545).

In the case of Bengali migration to the CHT in Bangladesh, poverty and adverse effects of climate change forced people to migrate, leaving their original place of land. However, other studies showed that political factors such as fear and repression in the plains region of Bangladesh have implications which force some people to migrate from their land (Siraj & Bal, 2017). After 1975, religious minority Hindu people experienced persecution and harassment from some Muslim people in many locations in Bangladesh. Thus, many Hindu people were forced to migrate to the neighbouring country of India as their safe place to live (Datta, 2004). In this connection, some Bengali Hindus may have migrated to the CHT to escape from torture, harassment, and sexual violence in their place of origin.

Interviews with climate migrants offer more detailed information on how climatic events and poverty are interlinked for their migration to the urban areas and the CHT in Bangladesh. Frequent climatic events (e.g., annual floods) damaged their livelihood and living options, and left nothing for the respondents in their place of origin to live on. Some interviewees who are now living in the urban slums mentioned that they had been affected by floods and cyclones repeatedly, and their entire region had been inundated and destroyed. Despite trying their best to adapt to the changing environment by setting up their dwelling in a nearby location to reduce their climate vulnerability, they had not been able to stay. A climate migrant now living in

Figure 4.1 Trend of climate migrants in Dhaka city

Source: Rana and Ilina (2021: 4).

a slum in Dhaka narrated that "due to the frequent inundation of his houses by the saline water, his family suffered terribly to manage livelihood and shelter. In one stage, he moved to Dhaka with family" (Climate migrant 14). This is the story of many of the climate migrants who live in the slums in Dhaka. The saline water intrusion, riverbank erosion, cyclones, and floods affected people in the rural and coastal areas in Bangladesh who live under serious risks and vulnerabilities (Dalsgaard & Ahsan, 2021).

Climate change impacts have forceful effects on the migration of the Bengali people to the CHT. Key informants mentioned poverty and government help as the determining factors in their migration. Although they had not been directly affected by climate change or environmental causes, they suffered secondary impacts in the form of diminished income and work opportunities. One climate migrant, a 50-year-old woman who migrated from the Faridpur district, explained how she came to the CHT and under what circumstances:

Researcher: Can you please tell me what happened in a place where you have come from?

Climate migrant 5: I had a big family in Faridpur consisting of three daughters and three sons. What I managed to earn per day was insufficient to feed my children one full meal per day. The rest of the day we remained hungry. In such a condition, the river took away everything we had. My husband moved to the *char* with us and built a *dochala*.[1] However, this place was damaged by river bank erosion. Our house and all our belongings were destroyed four times by river bank erosion.

Researcher: What did you do then?

Climate migrant 5: My husband and I worked hard to overcome the damage caused every time. My husband used to work as a day labourer, and I worked as a water carrier at a small hotel in the *bazar*. We also sought some help from the government but we did not receive any help.

Researcher: Why did you come to the CHT?

Climate migrant 5: We first went to Dhaka. However, life in Dhaka was tough. We had to pay money as extortion to the *Mastans* for hiring a house in the slum. But I did not feel safe in the slum. My daughters were not safe. Then, I heard that people are going to the CHT and the government is giving land and food. I decided to move. My husband and I first came to Chittagong and stayed for a few days. Finally, we came to the Rangamati district.

This account shows that the migration decision was taken only after the family had been affected repeatedly by climatic events that disrupted their livelihood and destroyed their home. However, the city as a migration destination is not safe due to the lack of facilities such as water, health and education, and law and order. Many informal settlement areas are governed by *Mastans*, a Bengali term for outlaw people. They control a particular territory through political connections and violence, demand money, and harass poor people who live in the slums (Jensen & Andersen, 2017). Newcomers are particularly prone to falling victim to *Mastans* because they arrive in a vulnerable position having run out of options and resources. Thus, some of the respondents share that they decided to migrate to the CHT to avoid the challenges posed by a precarious life in the city and follow the promise of land and food. Another climate migrant, a 45-year-old respondent originally from Sylhet, told a similar story about floods washing away his assets and forcing him into extreme poverty.

> I used to live in a rural area of the Sylhet district. I had a small piece of land for agriculture. With the land, I managed to sustain my family for about six months. The remaining six months I used to work on the land of other people as a day laborer. I faced devastating floods twice that destroyed everything. In the course of time, I fell into extreme poverty. I was fortunate that I got the news of the CHT. When I got the news from my relatives, I came here without hesitation.
>
> (Climate migrant 9)

The preceding accounts reveal that climatic events and poverty are connected in many places in Bangladesh. Climatic events have the potential to destroy and diminish the limited resources of people who are already poor and further impoverish them. The respondents already lived in poverty when floods destroyed their crops, and over time, their capacity and options to adapt were exhausted. Frequent climatic events (floods and cyclone) and poverty represent a threat multiplier for poor people in low land and rural areas in Bangladesh (Alam, Alam, Mushtaq, & Clarke, 2017; Islam & Hasan, 2016), eventually leaving them with no other option than to migrate to any suitable place. In this context, migration is seen as the "last resort" to save their lives (Penning-Rowsell et al., 2013).

Lack of income in the place of origin, and the hope for more income elsewhere, are related to the poverty situation of the respondents. In some areas of Bangladesh, such as Mymensingh and north Bengal, many people remain chronically poor. They lack minimum basic needs, such as income, daily consumption of food, medical facilities, clothes, and work opportunities (Sen & Hulme, 2006; Rahman, Matsui, & Ikemoto, 2008). In the 1980s and 1990s, these areas were more affected by poverty compared to other regions in Bangladesh (Rahman, Mahmud, & Haque 1988). Some of the poverty-stricken people migrated to the CHT after hearing about the opportunities for the land provided by the government. This means that the migration of the Bengali people to the CHT is also related to socioeconomic factors.

The preceding discussion presents the interplay of migration decision of the people with the association of push and pull factors. Climate change affected people try to avoid the vulnerabilities and migrate to the urban areas and other places where they find livelihood opportunities. The opportunities in the urban areas and CHT acted as the pull factors that attracted the poor people to come and settle.

Socioeconomic and political factors

The socioeconomic and political situation in their place of origin such as income, assets, and other factors such as debt, extortion, violence, and any threats that are commonly linked to chronic poverty constitute push factors in the migration decision to the urban areas and CHT in Bangladesh. A study conducted by Islam, Schech, and Saikia (2021:7) showed that more than 70 percent of the Bengali respondents who migrated to the CHT suffered from low income and insufficient land in their place of origin. In rural areas of Bangladesh, land ownership is directly connected to income. As discussed in Chapter 3, the country is one of the most densely populated countries in the world, and the cultivated area is on average about 0.125 acre per person (Quasem, 2011). The increasing population has led to fragmentation and reduced productivity of the land (Rahman & Rahman, 2009). Due to economic hardship, some respondents have become indebted. Poor people obtain loans from micro-credit banks in order to escape from the vulnerabilities and poverty, but then fail to repay the loans. This complex situation of poverty and indebtedness influenced the migration decision of some respondents (Afsar, 2004; Findlay & Geddes, 2011). Very few respondents mention incidents of physical violence, threats to family, and extortion as factors that pushed them to migrate.

Interviews with experts provided different views about the causes of migration of the Bengali people to urban areas and the CHT. For example, one expert, a university professor and director of a migration research center, acknowledged that poverty and environmental causes played a role in the migration to urban areas such as Dhaka and Chottagram. He argued that migration in an urban slum, particularly in Dhaka, occurs for the poverty and adverse effects of climatic events. While asking about the migration of the Bengali people to the CHT, the expert gave a different view about the migration to the CHT. According to the expert,

> I cannot say exactly that Bengali people who have migrated to the CHT are all climate affected. However, I can say that poor and landless people being hit by the riverbank erosion, floods and cyclone mostly moved to the CHT for land.
>
> (Expert 9)

Interviewees with expertise in the field of climate change issues also supported the view that a majority of Bengali migrants to the urban areas and the CHT had lost their homes, livelihood, and land due to floods, cyclones, and riverbank erosion in their place of origin. One professor in international relations and a CHT expert argued that the environment- and climate-affected landless and poor people moved to the CHT when the government announced that each family would receive land and livelihood.

An expert working as the director in a research center argued that "people affected by the climatic events, particularly by the floods and riverbank erosion come to Dhaka and live in slums because of their poverty. The working opportunity in the garment industry and other small industries also attract the poor people to come and settle in Dhaka" (Informant 11). In regards to the migration to the CHT region expert 2, a history professor and activist for indigenous people's rights, offered the following account:

> Poor and landless people from the low-lying areas of Bangladesh moved to the CHT only for land and livelihood. When the government announced that each family would receive five acres of land and 40 kilograms of rice per month, poor people in the coastal belt were tempted by the offer. This is why many displaced people moved to the CHT and received land, sometimes captured the land and settled there. Now, these Bengali people are the king (powerful) in the CHT.
>
> (Expert 2)

In the context of CHT migration, the pull factors contributed significantly to receive the decision of migration. Expert 12 suggests that Bengali settlers have grabbed land, echoing arguments elsewhere that many of the settlers have illegally occupied the land in the CHT (Adnan, 2004; Adnan & Dastidar, 2011). This has enabled them to consolidate their economic and social position in the CHT to the point that Bengali people now are situated in dominant positions in many parts of the region. According to expert 2, it was the pull factor of the promise of land and government support that led many Bengalis to migrate to the CHT. As the expert explained, the motive of the Bangladesh state was to gain hegemony over the tribal people of the CHT by settling tens of thousands of Bengalis. The government and the local administration in the CHT played a significant role in managing this migration process. The political motive is acknowledged in the literature on the CHT conflict, which also identifies the role of the army. Literature highlighted that the army wanted to settle Bengali people to act as a human shield as well as balance the population between the Bengali and tribal people in the CHT (Mohsin, 1997, 2000; Levene, 1999; Chakma, 2010). According to this perspective, Bengali migration would have been much less during 1979–1990 without the enticements of land, food, and protection offered by the government.

In summary, environment and climate change impacts influenced the poor people to migrate to the urban areas and the CHT to escape from climate vulnerabilities and poverty. One point emerging from the climate migrant interviews is that their migration to the CHT occurred in several steps, via Dhaka and other destinations. This indicates that the CHT was not the place of first choice. The availability of land and forest resources equally motivated respondents to migrate to the CHT. Although some experts highlighted the government's intention of establishing hegemony in the CHT as the main driving force behind the migration and settlement of the Bengali people to the CHT, other experts acknowledged the role of climatic events and poverty as causal factors in the migration and settlement of lowland Bengali people to the Dhaka and the CHT. Thus, the micro-level information from the climate migrants and interview information from the experts confirms that environment and climatic events, along with poverty in the place of origin, are the dominant factors for migration of the people to the urban areas and the CHT in Bangladesh.

The impacts of climatic events in migration decisions

The foregoing sections highlight that poverty and climate impacts played a significant role in migration decisions of the people. This section, thus, explores what climatic events pushed people to leave their place of origin. Different studies suggest that a majority of the migrants living in the slums had experienced floods, cyclones, riverbank erosion, and coastal erosion (Islam, Schech, & Saikia, 2021; Abir & Xu, 2019). However, the most frequent and devastating of climatic events afflicting the low-lying areas of Bangladesh are floods and riverbank erosion. Sea-level rise and drought also constitute a factor in displacement and migration. Interviews with climate migrants also provided information on how climate events affected their lives in their place of origin. One climate migrant, a 50-year-old man from the Barisal district, a southern part of Bangladesh, told about the measures he took to adapt to the frequent flooding of his home:

> I came to the CHT with my wife and five children through migrating Dhaka. In my place, we were affected by the floods, cyclone and intrusion of the sea-water. Our house was located in a lowland area. During the rainy season, flood water entered our house. We then built *macha*[2] (top place) on our house. In some cases, we could not leave the house for several days. We used a bamboo-made boat to go to work.
>
> (Climate migrant 4)

People in the low-lying areas of Bangladesh, such as Barisal, Noakhali, Bhola, Patuakhali, Khulna, Barguna, and Cox's Bazar, are more affected by floods and cyclones. The devastating cyclone Ayla in 2009, and cyclones Nargis in 2008 and Sidr in 2007, destroyed many houses, trees, roads, and agricultural fields in these areas. A 45-year-old man from Bhola district explained how his work as a fisherman became impossible due to the increasing severity of coastal cyclones:

> Poor people like me cannot live in such a place. Every year, our house has been destroyed. I used to work as a fisherman. I lost my nets and boat several times in a cyclone. I also shifted my house several times. In the end, I took shelter on the embankment of the river. It was a hard life, and my children had to starve many times. Life was good for me and my family when I managed to catch more fish from the sea. However, fishing at sea was risky. My boat was not good enough to face the strong winds and cyclones. Finally, I moved to Dhaka for escaping the vulnerabilities.
>
> (Climate migrant 11)

As Chapter 3 has outlined, coastal areas in Bangladesh are increasingly affected by sea-level rise and coastal erosion. This climate migrant's story is similar to that of many people in the lowland and rural areas who have become the victims of climatic events, enduring the loss of their livelihood, assets, and sometimes also loss of life (Ahmed, 2006; Brouwer et al., 2007; Mirza, 2003). In this situation, migration becomes an option of last resort to escape climate change–induced vulnerabilities and risks (Penning-Rowsell et al., 2013).

The climatic events damage the crops, house, and agricultural land in the rural and coastal areas every year. In an agriculture-based country such as Bangladesh, damage to crops causes crop failure and throws people into poverty (Muqtada, 1981; Gray & Mueller, 2012). Damage to their homes and land renders the affected people homeless and landless, eventually forcing them to be displaced. In addition, the loss of basic utilities such as fresh water, and damage to roads and infrastructure due to frequent climatic events, negatively affect people's quality of life. In some cases, family members of the respondents were injured or lost their lives during the events.

The interviews with climate migrants provided detail of climatic events and the consequences they experienced. A 45-year-old male migrant from Chittagong district narrated his experience of a devastating flood:

Researcher: Can you please tell me what exactly happened to you?

Climate migrant 6: One night I heard a sound '*shoo, shoo.*'[3] We were in our house. We could not understand it. However, suddenly we heard the announcement that floods are coming. We did not have time to go to a safe place. Suddenly, the flood water came to our house, with great force. Within a few minutes, the flood water destroyed everything. The flood water washed away all of our villages. I lost two of my brothers. I lost all my cattle and all our belongings with my house (Climate migrant 6).

As this interviewee pointed out, sudden climatic events are frequently devastating. Floods, cyclones, and riverbank erosion can destroy vast areas within the space of a few minutes. On the other hand, events such as coastal erosion and sea-level rise gradually destroy the livelihood and living options of people. Another climate migrant from the Barisal district and now living in the slum in Dhaka described his experience with sea-level rise and coastal erosion:

> Like me, many people have come to the slum in Dhaka. My previous home was very close to sea level. Every year, the sea-water entered our house and land. Cultivating in the salty water was impossible. We were not even able to get fresh drinking water. We spent several years living like this. Our health condition was also affected by the salty water and continuous waterlogging. Because of all these difficulties we moved to Dhaka.
>
> (Climate migrant 12)

People living in places close to the sea and big rivers are affected by frequent floods, cyclones, and coastal erosion. These climatic events frequently hit those areas and damage agriculture, trees, water sources, and cattle. People gradually lose land and the financial capability to lead their life. People are also deprived of the minimum medical facilities due to the lack of hospitals as well as many health professionals showing less interest in working or living in highly climate affected areas. Thus, climate affected people suffer from ill health because of climatic events and lack of health facilities. In such conditions, many people have difficulty managing their livelihood and other basic needs.

The connection between climatic events and displacement and migration is not inevitable. An early and appropriate response and practical assistance from the government during and after adverse climatic events may support people in their attempts to stay in or return to their homes. During and after floods, cyclones, and other environmental disasters, the timely response of the government is crucial in saving the life and property of residents. People who abandon their homes tend to do so because they do not receive adequate help from the government and are unable to continue living in the affected area. Interview with the climate migrants living in slums in Dhaka and the CHT proved that they received very little help from the government to sustain in their place of origin. Most of the climate migrants interviewed for this study said that neither the central government nor the local government authority offered support to the affected people. Only one climate migrant from the Sylhet district reported that some assistance was initially offered:

> In the wake of the disaster, the government representative came to see us and gave us some emergency assistance, such as light refreshments and water. After that, they left us. We never found them again. We tried to stay in our place, but it was unsuitable for living and finding enough

> food. Therefore, I went to Dhaka for a work opportunity. I struggled there to find a house and work. Finally, I moved to CHT.
>
> (Climate migrant 9)

Another climate migrant narrated that after cyclone Ayla, his family lost everything. The local government authority gave them temporary shelter and food. However, it was insufficient to lead a big family consisting of seven members. Thus, they moved to Dhaka and hired a house in Korail Slum in Dhaka (Climate migrant 13). In the absence of government support after disasters, climate affected people from low-lying rural areas had little option but to move to urban centers in search of work (Afsar, 2004; Ahsan, Kellett, & Karuppannan, 2016). This is demonstrated in the preceding interview segment, where the participant first moved to Dhaka before migrating to the CHT to find an income sufficient to support his family.

Indeed, local level help during and after climatic events was absent in most cases. It is difficult for the government to provide food and shelter. Access to land in the rural areas (both ownership and lease) is vital for farmers to grow food grains and construct a dwelling for their families. It is usually the landless rural families who are most likely to migrate when the rural economy is unable to provide a livelihood (Afsar, 2004). In some cases, extended family helped the affected people. Two respondents (Climate migrant 9, from Sylhet, and Climate migrant 6, from the Chittagong district), reported that relatives sent them food in the aftermath of floods and cyclones.

This section has explored the effects of climatic events, the cost (nonfinancial) that climatic events imposed on the respondents, and the information about the assistance the government and local administration provided. The story that emerges from the information provided by the climate migrants and experts is that climatic events, such as floods and cyclones, in the lowland and rural areas were very costly in that they destroyed the material basis for their lives. A similar story is told in other academic studies on population displacement in Bangladesh, particularly in low-lying and coastal areas (Gray & Mueller, 2012; Mallick & Vogt, 2014; Feldman & Geisler, 2012). People living on coastal and *char* land are highly vulnerable to climate change–induced displacement and migration (Alam et al., 2017; Islam & Hasan, 2016). Thus, the affected people consequently preferred to migrate to the urban areas and the CHT as they did not have any option in the place of origin or supporting help from the government and local authorities. As there was the possibility of getting land and resources in the CHT, climate affected people migrated to that region.

Behavioural and other factors of migration

The literature on the climate change and migration relationship posits that social and behavioural factors also influence the migration decisions of the people in the affected places. Apart from climatic events, behavioural and social aspects, such as age, gender, human capital, satisfaction level, and networks play an important role in people's decisions to migrate (Kniveton, Smith, & Wood, 2011; Martin et al., 2014; McLeman, 2014). The level of satisfaction about their standard of living significantly influences a person's decision to live in a particular location. People migrate from one place to another for opportunities, income, security, and future development (Adger et al., 2002; Biermann & Boas, 2008; Curran, 2002). The availability of food, security, and good opportunities for their children motivates people to stay or to move from a place. The low rating for the place of origin correlates with the accounts of climate migrants in the previous section who highlighted the miserable life they were facing due to poverty and frequent floods, cyclones, and riverbank erosion. The respondents also stated that they used to live in coastal areas that have become unsuitable for living in the wake of cyclones and gradual sea-level rise. One climate migrant, a 55-year-old female from Khulna district, expressed a deep level of dissatisfaction:

> Life in my place was not good. The only reason is frequent floods and cyclones. I used to live in the embankment of the river. The flood and cyclone destroyed my house twice. Therefore, I took shelter in the shelter centre. However, I was not able to manage food for my family. My life was terrible there. My husband advised me to move to Khulna city. As my husband know how to do business, I am now fine in the CHT.
>
> (Climate migrant 2)

The poor living conditions and failure of their livelihood influenced the migration decision. The social and economic conditions of the people provide insight about how people have been leading their lives in their place of origin. Another climate migrant living in a slum in Dhaka outlines that his previous place was very terrible. His family suffered from the scarcity of water and sanitation. His income was very poor to send his children to school. Now, his family is living in a slum and getting able to buy water and sanitation facilities. His children are also getting education facilities (Climate migrant 14). Another climate migrants living in the slum narrated,

> I lost everything due to floods and cyclones. As I lived in the coastal region in the Khulna division, I have to resettle my house several times. In most of the cases, I took loans from the NGOs. It was very difficult to repay the loan meeting the daily expenses of my family. Finally, I left my place and came to Dhaka.
>
> (Climate migrant 17)

Indeed, people at the local level in Bangladesh depend on NGOs and alternative financial sources as poor people can hardly get a loan from government-owned financial institutions. Moreover, receiving a loan from a government bank is quite a difficult and lengthy process. Thus, people seek loans from NGOs in the crisis period. The NGOs impose high interest rates that impose a lot of hardship to the loan receivers. In some cases, the loan receivers migrate to the urban areas in order to escape from the NGOs. A climate migrant highlighted this complex issue:

> I have been affected by waterlogging for several years. I shifted my house in several times. For this shift, I borrowed money from the local NGOs. I used to pay the loan every month. But I failed to pay the interest of the loan sometimes. The NGO workers used to put pressure on me to pay the loan timely. I became mentally sick to thinking about the loan and meeting the household costs of my family. Finally, I left my place and settled in slum in Dhaka.
>
> (Climate migrant 18)

The results concerning the behavioral and social aspects of the migration of the Bengali people to the urban areas and the CHT reflect their situation of dissatisfaction in their place of origin, as well as their involvement in occupations that are very vulnerable to climate impacts (for example, farming). The frequent floods and cyclones more often increase the level of dissatisfaction among the poor and low-income people when they repeatedly lost their agriculture and livelihood options. The physical damage caused by disasters also motivated people to abandon their homes. According to migration studies, people consider their present situation and the possible benefits in the destination place in their decision-making (Reuveny 2008; McLeman 2014; Hugo 2013).

The working opportunity in the city centers (any kind of work, such as day labour, household workers, small vendors) attract the poor and climate affected people. The resources in the CHT, particularly the land, attracted the poor and climate affected people to come to the region and become owners of the land. The perceived availability of land and forest resources greatly motivated Bengali migrants to move to the CHT and made it an attractive destination for them. Whether migration is climate change induced or poverty-driven, people prefer to move to a fertile place rather than to barren or desert land. The perception of the CHT as a land of resources and opportunities also attracted some middle-class people, such as government officials, to the CHT with a view to becoming landowners (Adnan, 2008). Some of the government officials who were posted to the CHT and became permanent settlers represented a small proportion of survey participants in this study. For most respondents, however, the aggregated dissatisfaction due to climatic events and poverty were the key motivation to migrate to the CHT.

Discussion

The findings outlined previously revealed the various factors that impacted on the migration decisions of the people who have come to live in the city centers and the CHT. In the case of the CHT the impacts of climatic events and poverty played a prominent role, adding a new perspective to the current understanding of this migration as political and state-sponsored migration (Adnan, 2004; Mohsin, 2003). Hearing about the possibility of a better life in the CHT from government sources and social networks persuaded them to migrate to this region despite the civil unrest and conflict that was building there. Chapter 6 will probe more deeply into the connections between migration and the CHT conflict.

The interview findings are supported by literature that has analyzed the human impacts of both sudden and slow onset processes of climate change, manifested in devastating floods, cyclones, and riverbank erosion in the low-lying, coastal, and river basin areas of Bangladesh (Gray & Mueller, 2012; Islam & Hasan, 2016; Islam & Siddiqui, 2016; Saha, 2017). As people lose their homes, land, and livelihood, they become caught in a deepening cycle of poverty. This experience is captured in the interviews with climate migrants who explained how the frequency and multiple forms of climate events detrimentally affected their livelihood and intensified poverty in the rural areas. The slow onset processes, such as sea-level rise, coastal erosion, and drought, are also destructive in some areas as they gradually destroy agricultural land, water sources, and the lives and livelihood of the people in the coastal and northern parts of Bangladesh (Shahid et al., 2016; Szabo et al., 2016).

This complex connection between climatic events and existing socioeconomic vulnerabilities in the coastal belt of Bangladesh forced some respondents to shift their homes several times and attempt to find different sources of livelihood. But despite showing resilience and attachment to their homes, gradual and/or periodic climate events shattered their hopes and efforts to rebuild their lives. Most of the respondents mentioned their inability to lead the life they wanted due to lack of income and frequent damage by climatic events. This finding is supported by the other studies in Bangladesh. For example, Findlay and Geddes (2011: 150) argue that climate affected people only migrated when they had lost all their assets and exhausted their options. Climatic events (e.g., floods) enforce an additional burden on poor people who then find there is no option but to migrate to the cities.

This chapter demonstrated that poverty was a key factor in the migration decisions of the people, but the frequent climatic events aggravated and sometimes caused the poverty situation. Climatic events worked as a threat multiplier, reducing people's livelihood options and thereby intensifying their poverty. Penning-Rowsell et al. (2013: S44) have termed migration as the "last resort" of the climate affected people. This is reflected in the deep levels of dissatisfaction with their place of origin reported by many of the Bengali respondents who migrated to the CHT. Studies on the relationship between climate change and migration argue that people decide to migrate when their satisfaction level crosses the "threshold" of dissatisfaction (Hugo, 2011; Meze-Hausken, 2008; McLeman, 2018). The experience of frequent climatic events, loss of livelihood, and dislocation on a number of occasions generated the deep level of dissatisfaction and forced people to migrate longer distances for permanent settlement (Reuveny, 2007). High levels of dissatisfaction due to the frequent climatic events and poverty influence some people to consider migration as their best adaptation strategy. Thus, climate change–induced migration is viewed as an adaptation strategy. Siddiqui (2010: 8) argues that "instead of viewing migration as a threat and vulnerability, the government of Bangladesh as well as the global community should incorporate migration as an important adaptation strategy."

To understand why some Bengalis have moved to the urban areas and the CHT, it is important to consider also pull factors, such as the possibility of getting land and livelihood options. In the case of the CHT, the government's promise of land in the CHT motivated them to get benefits in the destination place. As Reuveny (2008: 3) argues, "climate affected people may consider both the costs and the potential benefits of migration and migrate only if the expected benefits outweigh the costs." For the respondents in this study, the availability of resources in the CHT and promises of support from the government helped them in their migration decisions. The expectation of land and food rations to support life in the new place nurtured their hopes of being able to continue living off the land. In the case of the urban slum, the pull factors are income opportunity and shelter in the slum. However, living in the slum is extremely competitive, and slum dwellers have to compromise both financial and human rights for managing the influential people.

The decision to migrate is influenced not only by economic calculations but also by a number of social and behavioural factors (Hugo, 2013; McLeman, 2014). The inability of the government to mitigate vulnerabilities sometimes accelerates population displacement and migration (Piguet, Pécoud, De Guchteneire, 2011; Warner, 2010). In the wake of climatic events, people need food, shelter, and protection in order to adapt to the climate impacts. The interview information shows that the government played a minimal role in mitigating the impacts of climate change to enable the affected people to stay at, or return to, their homes. Research has highlighted the significant role of community support, communication, and trust in mitigating the impacts of disasters (Patterson, Weil, & Patel, 2010), and the role of the institutions and government support to enhance the resilience capacity of the people in the face of adverse climatic events (Reid & Alam, 2017). However, the respondents experienced a lack of community support and cooperation during and after climatic events. Thus, migration to the urban areas and the CHT was also a consequence of the failure of government and local community to support people in their attempts to cope with and adapt to the impacts of climatic events, and enhance their resilience and capacity to remain in their place.

Strengthening resilience capacity encompasses improving livelihood options, imparting knowledge, and providing technical support to develop alternative options for living. The democratic transformation from military rule in 1991, the green revolution which increased rice production, and support from official development assistance have enabled successive governments to implement adaptation and mitigation policies in the most affected areas of Bangladesh, which has consequently helped people to survive on their land. Moreover, Bangladesh is one of the least developed countries (LDCs)[4] that have formulated a long-term climate change strategy, the Bangladesh Climate Change Strategy and Action Plan (MoEF, 2009), in order to mitigate climate effects and take adaptation measures. Two funds, one using government resources (BCCTF) and the other using donor resources (BCCRF), are helping affected people to address climate change effects (Rai, Huq, & Huq, 2014).

Conclusion

From the evidence presented here it can be concluded that population movement to the urban areas and the CHT over recent decades occurred due to multiple factors, but climatic events and poverty have played a crucial role in forcing people to leave their places of origin. In the context of climate change, forced migration is usually a combination of displacement, escaping hardship, and hope for a better life in another place (Laczko & Aghazarm, 2009; Piguet, 2008). Climatic events destroyed the livelihoods of people as well as eroded the capacity to adapt in many parts of Bangladesh. Lack of support from the government at the national and local level also play a part in this. Climate change is not a sudden or recent phenomenon in Bangladesh, and affected people have migrated to cities and other regions such as the CHT for several decades.

The migration of the Bengali people to the CHT is not an exceptional case in the sense that it reflects the same story of migration due to adverse effects, livelihood failure, and no hope in their place. However, the CHT has particular features as a migrant destination in that it is known to have underutilized land and forest resources, and, on the other hand, to be a place of cultural difference and conflict. Thus, respondents could imagine rebuilding their lives on new land, but they also responded to the assurance of government support. Hearing about the possibility of a better life in the CHT from government sources and social networks persuaded them to migrate to this region despite the civil unrest and conflict that was building there. By examining the role of climate change in migration decisions, this chapter adds a new perspective to the current portrayals of the migration to the CHT as mainly a political and state-sponsored migration (Adnan, 2004; Mohsin, 1997; Chakma, 2010).

Notes

1 *Dochala* is a local term used in some parts of Bangladesh to refer to a small house constructed with bamboo and tin. Some people also use jute sticks to build the wall of the house.

2 *Macha* is a local term which is used to describe a covered platform made of wood and bamboo on top of the house or next to the house to take refuge from flood water.
3 'Shoo, shoo' refers to the sound come from rushing water and strong wind.
4 In 2018 the government of Bangladesh claimed that the country had been promoted to the status of a lower-middle income country from least developed one. The World Bank has also endorsed Bangladesh as a lower-middle income country as the per-capita income is $1,314 (World Bank, 2018).

References

Abir, T. M., & Xu, X. (2019). Assessing the factors influencing migration decision of climate refugees in coastal areas of Bangladesh. *American Journal of Climate Change*, *8*(02), 190.

Adger, W. N., Kelly, P. M., Winkels, A., Huy, L. Q., & Locke, C. (2002). Migration, remittances, livelihood trajectories, and social resilience. *AMBIO: A Journal of the Human Environment*, *31*(4), 358–366. https://doi.org/10.1579/0044-7447-31.4.358

Adnan, S. (2004). *Migration land alienation and ethnic conflict: Causes of poverty in the Chittagong Hill Tracts of Bangladesh*. Dhaka: Research & Advisory Services.

Adnan, S. (2008). Contestations regarding identity, nationalism and citizenship during the struggles of the indigenous peoples of the Chittagong Hill Tracts of Bangladesh. *International Review of Modern Sociology*, 34 (1), 27–45.

Adnan, S., & Dastidar, R. (2011). *Alienation of the lands of indigenous peoples: In the Chittagong Hill Tracts of Bangladesh*. Copenhagen: International Work Group for Indigenous Affairs. https://www.southasia.ox.ac.uk/sites/default/files/southasia/documents/media/oxford_university_csasp_-_work_in_progress_paper_15_adnan_dastidar_alienation_of.pdf

Afsar, R. (2004). Dynamics of poverty, development and population mobility: The Bangladesh case. *Asia-Pacific Population Journal*, *19*(2), 69–91.

Ahmed, A. U. (2006). *Bangladesh climate change impacts and vulnerability: A synthesis*. Climate Change Cell, Department of Environment, Bangladesh. https://www.preventionweb.net/files/574_10370.pdf

Ahsan, R., Kellett, J., & Karuppannan, S. (2016). Climate migration and urban changes in Bangladesh. In R. Shaw, A. Rahman, A. Surjan, & G. Parvin (Eds.), *Urban disasters and resilience in Asia* (pp. 293–316). Amsterdam: Butterworth-Heinemann.

Alam, G. M., Alam, K., Mushtaq, S., & Clarke, M. L. (2017). Vulnerability to climatic change in riparian char and river-bank households in Bangladesh: Implication for policy, livelihoods and social development. *Ecological Indicators*, *72*, 23–32.

Biermann, F., & Boas, I. (2008). Protecting climate refugees: The case for a global protocol. *Environment: Science and Policy for Sustainable Development*, *50*(6), 8–17.

Black, R., Adger, W. N., Arnell, N. W., Dercon, S., Geddes, A., & Thomas, D. (2011). The effect of environmental change on human migration. *Global Environmental Change*, *21*(1), S3–S11. https://doi.org/10.1016/j.gloenvcha.2011.10.001

Brouwer, R., Akter, S., Brander, L., & Haque, E. (2007). Socioeconomic vulnerability and adaptation to environmental risk: A case study of climate change and flooding in Bangladesh. *Risk Analysis*, *27*(2), 313–326. https://doi.org/10.1111/j.1539-6924.2007.00884.x

Chakma, B. (2010). The post-colonial state and minorities: Ethnocide in the Chittagong Hill Tracts, Bangladesh, *Commonwealth & Comparative Politics*, *48*(3), 281–300. http://www.tandfonline.com/action/showCitFormats?doi=10.1080/14662043.2010.489746

Curran, S. (2002). Migration, social capital, and the environment: Considering migrant selectivity and networks in relation to coastal ecosystems. *Population and Development Review*, *28*, 89–125.

Dalsgaard, J. L. T., & Ahsan, R. (2021). From the underwater to city slums. In F. W. Leal, J. Luetz, & D. Ayal (Eds.), *Handbook of climate change management*. Cham: Springer. https://doi.org/10.1007/978-3-030-22759-3_144-1

Datta, P. (2004). Push-pull factors of undocumented migration from Bangladesh to West Bengal: A perception study. *The Qualitative Report*, *9*(2), 335–358.

Environmental Justice Foundation. (2018). On the frontlines: Climate change in Bangladesh. https://ejfoundation.org/index.php?p=reports/on-the-frontlines-climate-change-in-bangladesh

Feldman, S., & Geisler, C. (2012). Land expropriation and displacement in Bangladesh. *The Journal of Peasant Studies*, *39*(3–4), 971–993. https://doi.org/10.1080/03066150.2012.661719

Findlay, A., & Geddes, A. (2011). Critical views on the relationship between climate change and migration: Some insights from the experience of Bangladesh. In E. Piguet, A. Pécoud, & P. de Guchteneire (Eds.), *Migration and climate change* (pp. 138–159). Paris: UNESCO Publishing.

Gray, C. L., & Mueller, V. (2012). Natural disasters and population mobility in Bangladesh. *Proceedings of the National Academy of Sciences*, *109*(16), 6000–6005.

Hafiz, M. A., & Islam, N. (1993). *Environmental degradation and intra/interstate conflicts in Bangladesh*. ETH Zurich: Center for Security Studies (CSS).

Hugo, G. (2011). Future demographic change and its interactions with migration and climate change. *Global Environmental Change*, *21*(1), S21–S33.

Hugo, G. (2013). *Migration and climate change*. Cheltenham, UK: Edward Elgar Publishing.

Islam, M. R., & Hasan, M. (2016). Climate-induced human displacement: A case study of cyclone Aila in the south-west coastal region of Bangladesh. *Natural Hazards*, *81*(2), 1051–1071.

Islam, M. T., & Siddiqui, T. (2016). Migratory flows in Bangladesh in the age of climate change: Sensitivity, patterns and challenges. In G. Wahlers (Eds.), *Panorama: Insights into Asian and European affairs* (Vol. I, pp. 49–65). Singapore: Konrad-Adenauer-Stiftung.

Islam, R., Schech, S., & Saikia, U. (2020). Climate change events in the Bengali migration to the Chittagong Hill Tracts (CHT) in Bangladesh. *Climate and Development*, 1–11. https://doi.org/10.1080/17565529.2020.1780191

Islam, R., Schech, S., & Saikia, U. (2021). Climate change events in the Bengali migration to the Chittagong Hill Tracts (CHT) in Bangladesh. *Climate and Development*, *13*(5), 375–385.

Jensen, S., & Andersen, M. K. (Eds.) (2017). *Corruption and torture: Violent exchange and the policing of the urban poor*. (1st ed.), Aalborg: Aalborg University Press. http://vbn.aau.dk/files/268209579/Corruption_and_Torture_online.pdf#page=120

Joarder, M. A. M., & Miller, P. W. (2013). Factors affecting whether environmental migration is temporary or permanent: Evidence from Bangladesh. *Global Environmental Change*, *23*(6), 1511–1524. https://doi.org/10.1016/j.gloenvcha.2013.07.026

Kniveton, D., Smith, C., & Wood, S. (2011). Agent-based model simulations of future changes in migration flows for Burkina Faso. *Global Environmental Change*, *21*(1), S34–S40. https://doi.org/10.1016/j.gloenvcha.2011.09.006

Laczko, F., & Aghazarm, C. (2009). *Migration, environment and climate change: Assessing the evidence*. Geneva: International Organization for Migration. http://publications.iom.int/system/files/pdf/migration_and_environment.pdf

Lee, S.-W. (2001). *Environment matters: Conflicts, refugees & international relations*. Seoul and Tokyo: World Human Development Institute Press.

Levene, M. (1999). The Chittagong Hill Tracts: A case study in the political economy of creeping genocide. *Third World Quarterly*, *20*(2), 339–369.

Mallick, B., & Vogt, J. (2012). Cyclone, coastal society and migration: Empirical evidence from Bangladesh. *International Development Planning Review*, *34*(3), 217–240. https://doi.org/10.3828/idpr.2012.16

Mallick, B., & Vogt, J. (2014). Population displacement after cyclone and its consequences: Empirical evidence from coastal Bangladesh. *Natural Hazards*, *73*(2), 191–212. https://doi.org/10.1007/s11069-013-0803-y

Martin, M., Billah, M., Siddiqui, T., Abrar, C., Black, R., & Kniveton, D. (2014). Climate-related migration in rural Bangladesh: A behavioural model. *Population and Environment*, *36*(1), 85–110. https://doi.org/10.1007/s11111-014-0207-2

McLeman, R. (2018). Thresholds in climate migration. *Population and Environment*, *39*(4), 319–338. https://doi.org/10.1007/s11111-017-0290-2

McLeman, R. A. (2014). *Climate and human migration: Past experiences, future challenges*. New York: Cambridge University Press.

Meze-Hausken, E. (2008). On the (im-) possibilities of defining human climate thresholds. *Climatic Change*, *89*(3–4), 299–324. https://doi.org/10.1007/s10584-007-9392-7

Mirza, M. M. Q. (2003). Climate change and extreme weather events: Can developing countries adapt? *Climate Policy*, *3*(3), 233–248. https://doi.org/10.1016/S1469-3062(03)00052-4

MoEF. (2009). Bangladesh Climate Change Strategy and Action Plan 2009, Ministry of Environment and Forests, Government of the People's Republic of Bangladesh, Dhaka, Bangladesh, xviii + 76pp.

Mohsin, A. (1997). *The politics of nationalism: The case of the Chittagong Hill Tracts*. Dhaka, Bangladesh: University Press Limited (UPL).

Mohsin, A. (2000). Identity, politics and hegemony. *Identity, Culture and Politics*, *1*(1). http://calternatives.org/resource/pdf/IDENTITY,%20POLITICS%20AND%20HEGEMONY-The%20Chittagong%20Hill%20Tracts,%20Bangladesh.pdf

Mohsin, A. (2003). *The Chittagong Hill Tracts, Bangladesh: On the difficult road to peace*. Boulder, CO: Lynne Rienner.

Muqtada, M. (1981). Poverty and famines in Bangladesh. *The Bangladesh Development Studies*, *9*(1), 1–34.

Patterson, O., Weil, F., & Patel, K. (2010). The role of community in disaster response: Conceptual models. *Population Research and Policy Review*, *29*(2), 127–141. https://doi.org/10.1007/s11113-009-9133-x

Penning-Rowsell, E. C., Sultana, P., & Thompson, P. M. (2013). The "last resort"? Population movement in response to climate-related hazards in Bangladesh. *Environmental Science & Policy*, *27*, S44–S59. https://doi.org/10.1016/j.envsci.2012.03.009

Piguet, E. (2008). *Climate change and forced migration. New Issues in Refugee Research, Research Paper No. 153, Policy Development and Evaluation Service, United Nations High Commissioner for Refugees*. Geneva: Switzerland. http://www.unhcr.org/en-au/research/working/47a316182/climate-change-forced-migration-etienne-piguet.html

Piguet, E., Pécoud, A., & De Guchteneire, P. (2011). Migration and climate change: An overview. *Refugee Survey Quarterly*, *30*(3), 1–23. https://doi.org/10.1093/rsq/hdr006

Quasem, M. A. (2011). Conversion of agricultural land to non-agricultural uses in Bangladesh: Extent and determinants. *The Bangladesh Development Studies*, *34*(1), 59–85.

Rahman, A., Mahmud, S., & Haque, T. (1988). *A Critical review of the poverty situation in Bangladesh in the eighties*. Dhaka: Bangladesh Institute of Development Studies, (BIDS).

Rahman, P. M. M., Matsui, N., & Ikemoto, Y. (2008). *The chronically poor in rural Bangladesh: Livelihood constraints and capabilities*. Florence: Taylor and Francis.

Rai, N., Huq, S., & Huq, M. J. (2014). Climate resilient planning in Bangladesh: A review of progress and early experiences of moving from planning to implementation. *Development in Practice*, *24*(4), 527–543. http://dx.doi.org/10.1080/09614524.2014.977761

Rana, M. M. P., & Ilina, I. N. (2021, December). Climate change and migration impacts on cities: Lessons from Bangladesh. *Environmental Challenges*, 5, 100242. https://doi.org/10.1016/j.envc.2021.100242

Reid, H., & Shafiqul Alam, S. (2017). Ecosystem-based approaches to adaptation: Evidence from two sites in Bangladesh. *Climate and Development*, *9*(6), 518–536.

Reuveny, R. (2007). Climate change-induced migration and violent conflict. *Political Geography*, *26*(6), 656–673. https://doi.org/10.1016/j.polgeo.2007.05.001

Reuveny, R. (2008). Ecomigration and violent conflict: Case studies and public policy implications. *Human Ecology*, *36*(1), 1–13. https://doi.org/10.1007/s10745-007-9142-5

Saha, S. K. (2017). Cyclone Aila, livelihood stress, and migration: Empirical evidence from coastal Bangladesh. *Disasters*, *41*(3), 505–526. https://doi.org/10.1111/disa.12214

Sen, B., & Hulme, D. (2006). *Chronic poverty in Bangladesh: Tales of ascent, descent, marginality and persistence*. Dhaka, Bangladesh: Bangladesh Institute of Development Studies.

Shahid, S., Wang, X.-J., Harun, S. B., Shamsudin, S. B., Ismail, T., & Minhans, A. (2016). Climate variability and changes in the major cities of Bangladesh: Observations, possible impacts and adaptation. *Regional Environmental Change*, *16*(2), 459–471.

Siddiqui, T. (2010). *Impact of climate change: Migration as one of the adaptation strategies* (Working paper series no. 18). Dhaka: RMMRU. http://www.rmmru.org/newsite/wp-content/uploads/2013/07/workingpaper18.pdf

Siddiqui, T., & Billah, M. (2014). Adaptation to climate change in Bangladesh: Migration the missing link. In S. Vachani & J. Usmani (Eds.), *Adaptation to climate change in Asia* (pp. 117–141). Cheltenham: Edward Elgar. https://doi.org/10.4337/9781781954737

Siraj, N., & Bal, E. (2017). 'Hunger has brought us into this jungle': Understanding mobility and immobility of Bengali immigrants in the Chittagong Hills of Bangladesh. *Social Identities*, *23*(4), 396–412.

Szabo, S., Hossain, M. S., Adger, W. N., Matthews, Z., Ahmed, S., Lázár, A. N., & Ahmad, S. (2016). Soil salinity, household wealth and food insecurity in tropical deltas: Evidence from south-west coast of Bangladesh. *Sustainability Science*, *11*(3), 411–421.

Warner, K. (2010). Global environmental change and migration: Governance challenges. *Global Environmental Change*, *20*(3), 402–413. https://doi.org/10.1016/j.gloenvcha.2009.12.001

5 Mediating factors in the migration process in Bangladesh

Introduction

> Migration options, and migrant agency are shaped and constrained by culture, economic, political, and social factors that operate at multiple scales, often well beyond the direct control or influences of the individuals, household, or local population.
>
> (McLeman, 2014: 230)

Migration and settlement of people from one place to another is not an easy process due to the contextual situation of the migrants and the destination place. The climatic events destroy all possible resources and thus limited the capacity to undertake a lengthy migration process. Second, destination places such as Dhaka are already overcrowded and lack resources and services. On the other hand, the CHT is remote and hilly, an almost separate region from other parts of Bangladesh (see Map 1.1), and inhabited by people who are ethnically, linguistically, and culturally distinct from other Bengali people. Furthermore, the region experienced an ethnopolitical conflict after 1975 which is also an issue of discouragement for the Bengali people to migrate to the region. Nevertheless, the post-1975 period witnessed a massive Bengali migration to the CHT. In such situations, mediating factors play an important role in enabling the migration of Bengali people from various parts of Bangladesh.

As discussed in the literature review chapter, migration is not just the result of the interplay between push and pull factors but also shaped by mediating factors, which often play a crucial role in the process of migration. Most migrations involve various drivers that can be deterministic, proximate, intervening, or mediating. As deterministic drivers, natural disasters, famine, and political conflicts force people to migrate (Warner, 2010; Marino & Lazrus, 2015). The expectation of a good life, the possibility of resources, income, safety, and security act as proximate causes (also known as pull factors) for both poor and well-off people to migrate (Van Hear, Bakewell, Long, 2018). In addition to these deterministic and proximate causes,

DOI: 10.4324/9781003430629-5

mediating factors such as networks, connections, and support enable aspirant people to migrate even to an adverse location (McLeman, 2014; Hugo, 2013; Reuveny, 2008). In the case of migration and settlement of people in Bangladesh, the mediating factors are the history of Bengali migration, political and military support, and social networks that facilitate the migration process.

As discussed in the contextual study chapter migration of Bengali people from the rural to urban migration as well as migration from the plainland to the CHT has a long history. The sheltering of poor and environmental change–induced migrants in the urban areas has a long history and a rather normal process. People used to come to the city center for work opportunities. This trend has increased significantly after the 1980s with the rapid urbanization and industrialization. Now, the situation is acute as more and more climate change-induced migrants are coming to the major cities every year for their shelter and livelihood. This migration process is mostly organized by the connections and social networks of the migrants. But in the CHT the British and Pakistani colonial powers promoted the migration process for their economic benefit and to gain political control over the area (Mohsin, 1997; Adnan, 2004; Tripura, 2008). After 1975, the government of Bangladesh continued the policy that was initiated under Pakistani rule and encouraged migration to the CHT to gain political control as well as reduce population pressure in low-lying areas of Bangladesh (Lee, 1997, 2001; Suhrke, 1997). This significant increase in migration flows during the period 1977–1989 has been categorized as "political migration," which exacerbated the emerging ethnopolitical conflict (Adnan & Dastidar, 2011; Chakma, 2010; Mohsin, 1997; Panday & Jamil, 2009). This migration process was sponsored by the Bangladesh state and specifically intended to diminish and weaken the aspirations of the tribal people to the right to self-determination in the CHT (Mohsin, 1997; Adnan & Dastidar, 2011).

The growing presence of the army in the CHT also played a mediating role in enabling Bengali settlement during 1977–1989 (Mohsin, 1997; Adnan, 2004; Chakma, 2010). In addition to state-sponsored migration, many people have also migrated of their own accord following the migration pathways and networks established by previous migrants. Thus, it is observed that family ties, networks, and communication played a role as general mediating factors to influence other people to migrate. But the government migration scheme and direct support by the local administration and security forces were the specific mediating factors for various reasons such as control of the CHT, assisting poor Bengalis, and reducing the population pressure in the lowland and rural areas.

This chapter critically analyses the mediating factors of the migration process in Bangladesh. By discussing the mediating factors, this chapter offers insights into how the migration process was institutionalized in a complex, overcrowded urban city and a conflict-prone location in the CHT, despite the people in the host place never welcoming the migration process. The chapter begins by discussing the historical roots of migration to the CHT that originate during colonial rule. An often repeated argument in the academic literature and in interviews with key informants is that colonial rulers facilitated Muslim Bengali settlement in the CHT in order to extract economic benefits from the region, as well as to keep control over the region. It is also shown that government assistance, army aid, family connection, and information flows helped them to migrate and settle in the CHT in the post-1971 period. As migration is a continuous process shaped by numerous and changing factors (McLeman, 2014), this chapter shows that migration is perceived to be still happening but possibly in diminished numbers, and now mainly with the direct help of family members and social networks.

The combination of historical links, family connections, kinship, information, and assistance from the government and army enabled Bengals to migrate to the CHT. Interviews with climate migrants have unveiled other factors beyond government help which assist people to migrate to the CHT. First, this section analyses the mediating factors in the migration based on the views of academics and key informants.

Academic and expert views about the mediating factors

Chapter 3 highlighted that the migration process in urban areas such as Dhaka is a natural process. Poor and climate-affected people come to the cities and settle for the livelihood. In the case of the CHT, the academic literature portrayal of the Bengali migration and settlement is as a political project of successive governments in Bangladesh. It is a view that is focused on the macro-level analysis of hegemonic construction and dominance over the tribal people in the CHT. What has received less attention from Bangladeshi academics is the role of migration history and how migrant networks and the government migration program facilitated the Bengali migration process. Moreover, in the Bangladesh period, there are various interpretations of the role of the state and the army in the Bengali migration. Some argue that the main purpose of the state sponsored migration program was to rehabilitate war refugees (Chakma, 2010: 290); others highlight the ways in which migration facilitated the state's maintenance of law and order in the CHT and protected the army in the CHT (Mohsin, 1997). Although there is no specific information on how migrant people protected the army in the CHT, the literature argues that massive Bengali migration to this region provided a justification and also the courage to the army in operating their missions. Moreover, on some occasions, Bengali settlers collaborated with the army and fought against the Shanti Bahini (Chakma, 2010; Mohsin, 1997). Eventually, the Bengali migration enabled "Bengali colonisation" (Mohsin, 2003: 33) and Islamization in the CHT (Mohsin, 2003; Panday & Jamil, 2009), but these processes are not separate phenomena; rather they are interconnected events in the Bengali migration and settlement to the CHT. This section discusses the academic literature and the views of the key informants, many of them academics, to present the ways history and politics are mediating factors in the Bengali migration and settlement to the CHT.

The historical connection has a role in furthering migration and settlement of the Bengali people in the CHT. It means that past migration works as a "chain migration" or "migration corridor" (Hugo, 2013) to influence more people to come and settle in the CHT. The migration corridor is used to analyze the international migration that denotes "an accumulation of migratory movements over time and provide a snapshot of how migration patterns have evolved into significant foreign-born populations in specific destination countries" (IOM 2018: 77). In the case of the urban migration and migration to the CHT, the migration corridor refers to the movement of the people from other parts of Bangladesh with the help of different actors such as government and networks. As mentioned in Chapter 3, people come to the cities by using their network and peer groups. As an expert highlights the process of urban migration,

> Migration of the climate displaced people to the urban city such as Dhaka has happened due to the adverse effects of climate changes and poverty. Initially, climate-affected people try to adapt to the situation. But they search for better options through social networks and family members who are already living in the urban cities. The work opportunities in the garments industry in this case encourages them to come and settle. By working in the garments factory, at least they get the scope to work in the garment industry, or they work as household workers.
>
> (Expert 11)

The climate displaced poor people settle in the urban slums with the help of the nonformal administration, local Mastans (i.e., local gang members), and sometimes by police. In such a case, the poor people are to give bribes or some money to access the basic services. In urban areas, services such as water, electricity, and sanitation facilities are scarce. Thus, slum people are to buy these services by giving money to the local Mastans and sometimes to the police.

But in the case of the CHT, the successive rulers from the British colonial period to the Bangladesh period encouraged Bengali settlement in the CHT to enable the extraction of economic benefits and maintain control of the region. The British, in 1933, annulled Rule 52 of the Regulation of 1900 Act (according to Rule 52, no nontribal people could enter and live in the CHT without the permission of the district commissioner (DC)) which opened the door for the Bengali people (Mohsin, 1997: 34). Thus to some experts in CHT affairs, migrant labour in the cotton fields marked the beginning of organized nontribal settlement in the region (Chatterjee, 1987; Tripura, 2008). One expert, a political leader from one of the CHT tribal groups, criticized British colonial rule and the tribal elite for their role in transforming the CHT from a tribally dominated region with special status to an integral part of Bangladesh with a mixed Bengali and tribal population. In response to a question about the beginning of Bengali migration to the CHT, the expert made this point:

> Migration of the Bengali people to the CHT is an old issue, but in the past only very few Bengali people lived in the CHT, usually for business and administrative purposes. But the British colonial rulers invited Bengali people to work in the cotton fields. Our tribal leaders did not oppose the decision of the British Raj, but instead helped the colonial ruler to extract the resources from the CHT.
>
> (Expert 5)

According to this expert, a very limited number of Bengali people lived in the CHT during the colonial period, and as noted in Chapter 3, only 2 percent of the CHT population in 1901 were nontribal people. But the colonial administration's agricultural development policy led to an increase in the number of Bengali migrants to the CHT, and this informant suggested that the tribal leaders agreed without realizing how this migration would eventually change their special status within the CHT. Adnan (2007: 4) described these non-Hill people inhabitants as "Bengalis comprising of government officials, security forces, professionals such as lawyers and teachers, as well as traders, shopkeepers, craftsmen, wage workers, rickshaw-pullers, and peasant cultivators." Several experts held the view that these migrants brought benefits to the region through their role in a reciprocal exchange system of goods and agricultural products between tribal and Bengali people in the CHT. One expert of indigenous background, a director of a rights-based organization, described this relationship as follows:

> In the British period, the tribal people maintained a good relationship with the Bengali people. Bengali people used to come, stayed overnight or sometimes for a longer period in the CHT. The Bengali business people and the trading system at that time benefited both the tribal and Bengali people. The tribal people used to sell bananas and other agricultural products to the Bengali traders and also bought goods. It built a reciprocal relationship in the CHT.
>
> (Expert 6)

The expert paints a picture of the past where Bengali people travelled and stayed in the CHT occasionally for business purposes. However, business from the British colonial period was eventually monopolised by Bengali people, and Bengalis living in the Chittagong districts captured all the business opportunities in the CHT (Shelly 1992: 81). These business and work connections with the CHT led to a gradual increase in the Bengali population, and by 1947, it was around 7 percent (8,762 persons) of the overall population of the region. By 1951, the proportion of Bengali residents had reached

9 percent (Chakma, 2010: 291; Adnan, 2004: 15). A decade later, the Pakistan government annulled the special status of the CHT as an "excluded area"[1] in 1963 and permitted the Bengali people to settle freely in the region (Mohsin, 1997: 46; Panday & Jamil, 2009). It is also alleged that the Pakistan government appointed Bengali officials in key positions of administration in the CHT and transferred the tribal officials to other parts of East Pakistan (Mohsin, 1997: 46). According to Mohsin, the aim of this administrative policy was to undermine the traditional institutions, reduce the influence of the local headmen, and weaken the position of the CHT as an excluded area. The government of Pakistan also implemented economic policies to promote national development in the region, for example, the development of the Kaptai Dam and Karnaphuli paper mills, which provided employment opportunities (Mohsin, 1997: 77). This encouraged many Muslim people from other parts of Pakistan to migrate to the CHT for work in the development projects (Mohsin, 1997: 106). Expert 3, a university professor and an expert on CHT affairs, offered his view of the consequences of such development projects:

> The development projects implemented by the Pakistan government disempowered the tribal people but benefited the Pakistan government. The projects (for example, Kaptai dam project) in one sense displaced many of the tribal people from their home and land but in another sense helped the Bengali people to come to the CHT to work in the projects.
>
> (Expert 3)

The tribal people showed little or no interest in working in these development projects as they were implemented against their will. As a result, the Pakistani rulers encouraged Bengali people to come and work, which also helped these migrants to improve their economic condition. As Mohsin and Ahmed (1996: 275–276) pointed out, "the construction of Karnaphuli Paper mill in 1953 displaced Marmas from the region and enabled the Bengali to settle. Now 100 percent Bengali population live in the adjacent area of Karnaphuli Paper Mill." The introduction of the Amendment of Rule 34[2] in 1964 further encouraged non-Hill people to settle permanently as it enabled them to own property if they had been living in the CHT for 15 years continuously (Ullah, Shamsuddoha, & Shahjahan, 2014: 201).

A key motive of retaining the special administrative status of the CHT and amendment of the Rule 34 policy was to control the non-Muslim CHT by encouraging more Bengali Muslim people to settle in the CHT. The Pakistan government regarded the tribal people with suspicion because during the partition of the Indian subcontinent in 1947 some tribal leaders wanted to be annexed with India instead of Pakistan (Mohsin, 1997: 35–36). This policy of encouraging Muslim Bengalis to migrate helped to increase the number of Bengali people who were deemed to be more supportive of the government in the CHT during the Pakistan period (1947–1971). Expert 2, a university history professor and expert in CHT affairs, argued that major changes in the constitution regarding the CHT affairs (retention of the status of the CHT and Rule 34) enabled the government of Pakistan to push its policy of Islamization into the tribal-dominated area through sending Bengali Muslim people to the CHT. When Pakistan emerged as a new state based on Muslim religious identity, the new government put a great deal of effort into consolidating Islamic values in every corner of the country, including the CHT inhabited by a majority of non-Muslim people (Mohsin, 1997). Indeed, after emerging as an independent state in 1956, Pakistan introduced Islamic law declaring the state an Islamic Republic of Pakistan. This declaration triggered the expansion of Muslim culture across the country (Choudhury, 1969). The CHT as a part of Pakistan experienced the same rule as was implemented in other parts of the country.

When Bangladesh became independent in 1971, its ruling class ignored the demands of tribal people for constitutional recognition of their cultural and political identity and urged them to assimilate into mainstream Bengali society and culture (Chakma, 2010; Mohsin, 1997: 58–59). The newly formed government also implemented a "rehabilitation policy for refugees displaced by the liberation war by resettling them to the CHT which in turn displaced some tribal people from their land" (Chakma, 2010: 290).

Change in the CHT accelerated after 1975 when a group of army officers killed the founding father of the nation, Sheikh Mujibur Rahman, and established a military government. The military government brought major changes in the CHT policies including large-scale Bengali migration to the CHT. With regard to this Bengali migration, Expert 2 argued that "Bengali settlement to the CHT was the desire of the army who have been struggling to maintain law and order." Some studies similarly suggest that based on the recommendation of the army, the Bangladesh military government adopted the policy of settling Bengali people with a view to suppressing the spirit of the self-rule movement (Mohsin, 1997; Levene, 1999). Another study argues that the government implemented the Bengali settlement policy to balance the ethnic composition of the population so that the Bengalis could act as "human shields" for the security forces (Adnan, 2007: 13). Hence Expert 7, a civil society person of indigenous background, suggested that migration and settlement of the Bengali people to the CHT was a "political decision" of successive governments. While there are different opinions about the nature of the political motivations, experts suggest that successive governments intended to increase the number of Bengali people in the CHT to weaken the movement based on tribal identity and gain greater control of the region.

The Bangladeshi academic literature and the experts' perspectives overlap in their political explanation of Bengali migration to the CHT which highlights the role of the state from the colonial period to the present. According to these accounts, the Bangladesh state played a mediating role by pursuing policies that encouraged Bengali people to settle in this culturally different region. The state intervened in the CHT for various reasons at various times and promoted the migration to the region to achieve its objectives. The objectives were different in nature based on the characteristics of the state. Whereas the British colonial power emphasized the economic interest and tax extraction, the Pakistan and Bangladesh states aimed to establish sovereignty and hegemonic control over the state territory.

Views of climate migrants about the mediating factors

This section investigates the perception of the respondents about the information, communication, actors, motivations, and channels that enabled their migration to the urban areas and the CHT. It sheds light on what the respondents considered as mediating factors, or sources of help and support while migrating to the cities and the CHT. As the previous chapter argued, the prospect of getting livelihood options in Dhaka, and land and resources in the CHT, encouraged people to migrate when they suffered from climate change events and lost everything in their place of origin. In order to fulfil these migration expectations, migrant people depended on their own networks and family connections, as well as direct help from the government and law-enforcing agencies.

The climate migrants living in the slums in Dhaka were asked what factors enabled them to come to Dhaka for settling. Most of the respondents highlighted the issue of work opportunities in urban areas. While asking about who helped them arrive to the cities, most of the respondents mentioned the presence of family connections and social networks. The already settled people in the slums initially help the poor climate-affected people to take shelter with them and search for their own houses. In the slums, hiring a house is also a difficult task. The local influential people and Mastans control the process and administration of the slum dwellers. One climate migrant mentioned that,

> I used to live in uncertainties and vulnerabilities as I lived in the coastal areas. Riverbank erosion was our constant fear. There was also a scarcity of livelihood and income opportunities in my place. I depended on my small piece of land and seasonal works. However, climatic events interrupted my livelihood options. My cousins used to live in Dhaka as garments workers. He helped me to settle in Dhaka when I left my place and came to Dhaka.
>
> (Climate migrant 17)

However, the migration process to the CHT region is quite different from urban migration. The respondents were asked to identify the most important reason for choosing the CHT as their migration destination. Most of the interviewees shared the views that they were encouraged by multiple factors such as the government's help to come to the CHT, the opportunity to get the land and other resources, family connections, and the purpose of business. Overall, this information suggests that the state played an intervening role in the settlement of Bengalis in the CHT. As shown in Chapter 4, many of the poverty and climate-affected people had no resources and options left and therefore welcomed the opportunity of migration to urban areas and the CHT, where they were promised land and livelihood options. Information, resources, and the supporting role of the government and already settled people play a key role in migrating to a new place (Van Hear et al., 2018; Hugo, 2013). The miserable life of the respondents in their climate affected places and poverty were the first conditions of migration, but the government's allocation of land to each family helped to set the migration process in motion and directed it to the CHT.

Some of the climate migrants stated that they were informed about the opportunity to migrate to CHT by representatives of local government, villagers, and family networks. The government's decision to encourage Bengali settlement and the possibility of getting land in the CHT has undoubtedly influenced the Bengali people to migrate to the CHT, confirming arguments about the state's role in the literature and key informant interviews. Literature review suggested that the army presence was critical to the survival of the Bengali community in the CHT. If the army camps are withdrawn from the CHT, Bengali settlers will not able to stay in the CHT. They might go back to their place of origin (Mohsin, 1997; Adnan, 2004). Academic research also describes the army as playing a vital role in facilitating the civil administration to settle Bengali people in the CHT (Mohsin, 1997: 174; Levene, 1999: 344).

Literature review show the role of the army in institutionalizing the Bengali migration in the CHT. However, the interviews with key informants offer some explanation for the reluctance of Bengali respondents to identify the role of the army. Academic studies support that the presence of the army and Bengali settlement in the CHT are connected. The relationship between the army and Bengali settlers has been described as "symbiotic" (Guhathakurta, 2012) whereby the presence of the Bengali settlers justifies the deployment of the army in the CHT and budget allocation to the military, ostensibly to maintain security in the region. The army presence gives Bengali settlers a sense of security and protection against potential or actual hostility from long-term residents (Mohsin, 1997, 2003; Chakma, 2010). Some climate migrants highlighted that they migrated from their place of origin and came to Chittagong city, where they registered their name as settlers in the army camp and finally came to the CHT. During the massive migration phase (1977–1989), a temporary camp was set up in the Chittagong region with the direct help of the Bangladesh army. After arrival in Chittagong, Bengali people stayed in the camp for enlisting their name and then were sent to different regions of the CHT (Siraj & Bal, 2017).

Most of the respondents were silent on the army role in the CHT and tended to avoid answering questions related to the army. In research, especially in conflicted environments, respondents may willingly misguide the researcher and avoid telling the truth due to fear, deprivation, and trauma (Goodhand, 2000; Höglund, 2011). The long-standing conflict, past trauma, and fear may hinder the survey respondent's willingness to expresses the true facts about their migration. In this situation, the respondents may have been reluctant to talk about the army because they are fearful of negative consequences. Mentioning any form of assistance or protection received from the army might be interpreted as going against the official discourse of the army playing a peacekeeping or neutral role. Besides the army, the civil administration (Office of the District Commissioner) in the CHT also played an active role in settling the Bengali people in the CHT. As a civil administration, the Office of the District Commissioner carried out the decision of the government to allocate land to the newly arrived Bengali people and ensure the implementation of the ration system, which provided new settlers with food grains.

In relation to other sources of assistance, climate migrant interviewees were more forthcoming. They came to know about the opportunities in Dhaka and the CHT from villagers or relatives who had already established themselves there. But the announcement of the government giving land and food rations to each family encouraged many of the people to settle in the CHT. Climate migrant 9, a 45-year-old male who came from the Sylhet district and lost all his belongings in the floods, explained how relatives and government assisted in his decision to migrate:

> I got the news from my relative that people are going to the CHT. The Government is giving land and food. As I struggled in my place of origin, I seized the opportunity. I came with my whole family.
>
> (Climate migrant 9)

The local authorities in the respondent's place of origin (members and leader of the Union Parishad) also disseminated the news of migration to the CHT. In some cases, local representatives were requested to spread the news of the Bengali settlement program in the CHT among the poor and climate affected people. For example, a 'Secret Memorandum' (Memo No. 665C) written to the Commissioner of the Chittagong Division, requested him "to collect particulars of the intending families from the Chairman of the concerned 'Union Parishad'" (Mohsin, 1997, appendix 1). Climate migrant 7, a 50-year-old male who came from the Bagerhat district to escape flooding and poverty, gave an account of local government involvement in his migration experience:

> One day the local chairman came to see us and observed that we were passing through a hard time. He informed me that the government was allocating land and rice if people were settling in the Hills area in the Chittagong. This put the idea in my head of going to Chittagong. After a long journey from my place we finally arrived there.
>
> (Climate migrant 7)

Local government networks were an important cog in the system of information distribution, instilling trust in the information and assuring potential migrants about practical government support. This was particularly important when migrants faced a long journey, as in this informant's case. In the CHT the local government helped to scrutinize the aspirant people who would like to go to the CHT as migrants (Siraj & Bal, 2017: 407).

In addition to the local government role, network and family connections are known to play an essential role in any form of migration in Bangladesh (Martin et al., 2014; Kuhn, 2003). The networks and family members not only inspire people to migrate but also help them in practical ways to move to a new place. In this process, a group of people in a village, and sometimes the majority of people in a village, have migrated to overcome their vulnerabilities and get access to a better life. In the case of the CHT, family connections and social networks among the Bengali people played a dual role: to facilitate migration and to make settlement more sustainable. In some cases, people who had already settled invited more people to come to the CHT to strengthen their position vis-à-vis the tribal inhabitants. As one climate migrant, a 50-year-old man from the Chittagong district, and now a leader of a local neighbourhood (*para*) in the CHT, explained:

> I first came to the area and registered my name with the administration. After registering the name, the authority (army/district commissioner) allocated me 5 acres of land. When I came to the land, I saw it was a hill and jungle. I built my house, but I could not live there due to the fear of the tribal people. I then invited some families in my local area place of origin. I informed them that there is enough land here, as well as tress. If you come here, I will take care of you. With my assurance, around ten families came and built their houses.
>
> (Climate migrant 3)

This suggests a particular type of chain migration where Bengali settlers encouraged neighbours and friends from their home village to settle in the CHT and defend their interests against the tribal people. In other cases, family connections worked as a force to motivate people to migrate to the CHT. The information, resources, and assistance provided by the community and government level enabled this migration process.

As a contextual issue, demographic pressure and livelihood failure also played an intervening role in migration decisions (Van Hear et al., 2018). Behavioural and social issues such as age, expectation, and hope for a better future also contribute a mediating role for the migration process (Martin et al., 2014). A study suggests that it was mostly young people aged 20–30 years who migrated to the CHT (Islam, Schech, & Saikia, 2021). Another study shows that young people generally come to Dhaka for pulling rickshaws and diversifying their income. This form of migration is a seasonal one where poor people come for a short period of time for their income diversification (Khandker, Khalily, & Samad, 2012). In the rural areas, the unemployment rate is very high in the lean season. Thus, young people migrate to the urban areas for the extra income.

Overall, it can be argued that mediating drivers strongly influenced people to migrate to urban areas and the CHT. In particular without social networks and family connections, the migration process would not have been possible at such a rapid pace during the period 1975–1989. Siraj and Bal (2017: 403) citing Adnan and Dastidar (2011) termed these people as the "self-propelled" Bengali migrants who have migrated from the adjacent areas of the CHT. The government and the army do not help the Bengali people to settle in the post Peace Accord period, but the "practice of nepotism, cronyism and favouritism" (Siraj & Bal, 2017: 403) by the administration and already settled Bengali people enable Bengali people to settle in the CHT. However, the massive migration depended mainly on the political decision of the government to not only actively encourage the idea of the CHT as a place of migration, but also provide material and in some cases logistical support for settlement. By providing land and food, the government alleviated the costs of migration to the migrants so that they could sustain themselves in their new location. The strong army presence also reduced the risks associated with settling in the CHT by ensuring the safety and security of the migrants. Thus, this section demonstrated that strong structural and political support was required to institutionalize migration to a remote and contested place such as the CHT. In addition, families and relatives helped to find work opportunities and livelihood options, and provided security in numbers. The already settled people assisted the new migrants. All these mediating factors worked together in facilitating the migration process.

Discussion

Mediating factors perform a crucial role in migration processes. In the sending and receiving places, these factors act as a conduit to facilitate the process of migration. Findings of this chapter support that mediating factors have played a significant role in addition to push and pull factors to accomplish the migration to urban areas and to the CHT. Mediating factors influenced the poor people as well as enabling them to settle in the slums in Dhaka and the CHT, and by doing so, these factors have also influenced the conflict situation in the region. This chapter has examined a number of mediating factors, namely the role of the government, security forces, migration policy, social networks, and family connections that shaped the migration process to the CHT. The mechanism of migration to the urban areas and the CHT can be aptly described in terms of the "push-pull plus" framework (Van Hear et al., 2018). According to the push-pull plus framework, migration is organized by the push factors (floods, poverty, war), pull factors (land, economic benefit), and additional factors such as networks, connections, migration history, and government policy (Van Hear et al., 2018).

The current study shows that most of the respondents came from low-lying and rural areas of Bangladesh where they were affected by floods, cyclones, and poverty. However, they did not necessarily migrate directly to the CHT. Interviews with climate migrants show that they first attempted to remain in their home area or migrated to urban centers to find employment. Over a period of time, information about opportunities in the CHT, including the prospect of obtaining land and help from the government, motivated them to migrate to the CHT. Thus, composite factors of environmental and demographic pressure, poverty, social networks, and political decisions by the government contributed to the migration of Bengali people to the CHT.

The role of mediating factors can be summarized in a simple graph (Figure 5.1), which is derived from Reuveny's (2008) framework. According to this framework, the contextual factors of the place of origin A push people to migrate to the destination place B. But the resources and opportunities in the place B also attract people. In this process of migration, other factors, which are called the mediating factors C, facilitate people's decision to move from A to B.

Applying this model to Bangladesh, the place of origin is usually a rural place in lowland Bangladesh, which has been affected by poverty and/or frequent climatic events, and the migration destination is the CHT, which is perceived by migrants as a place with a lower population, enough land and forest resources, but also as remote, ethnically different, and suffering from conflict (see Figure 5.1). Mediating factors play a crucial role in motivating people to migrate to the CHT, as well as institutionalizing the migration process in the face of local opposition and an unfamiliar environment.

As there is no literature on the Bengali migration to the CHT, this chapter has relied on what the international literature has to say about mediating factors (McLeman, 2014; Van Hear et al., 2018; Reuveny, 2008) in order to reassess the drivers of Bengali migration and settlement in the CHT. One important factor was government policy to reduce the population pressure in the rural areas of Bangladesh and giving livelihood options to the poor and

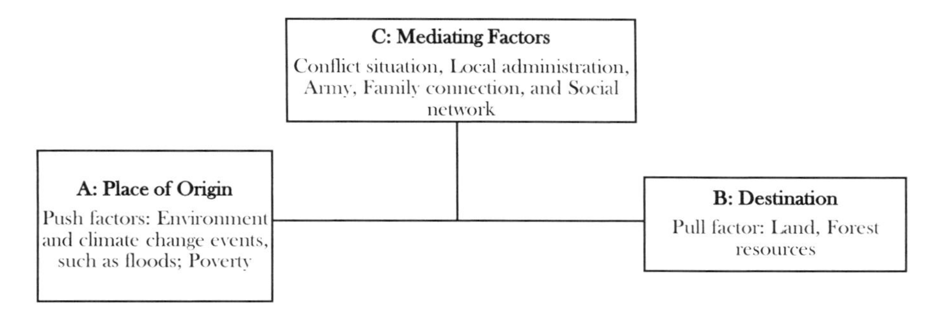

Figure 5.1 Mediating factors in Bengali migration to the CHT

Source: Developed by the researcher based on Reuveny (2008).

those people induced to move by environmental change. The army is another mediating factor because it played a direct role in politics and policy-making at the time of large-scale assisted migration (1977–1989). Successive military-backed governments headed by army generals during this time formulated the CHT policy to achieve a larger proportion of Bengalis in the CHT population and normalize the conflict situation of the CHT.

Migration decisions take into account risk, finance, connections, and easy communications (Van Hear et al., 2018); thus people tend to migrate to a place which they perceive is safe, accessible and lucrative (Lee, 1966). In that case, migration is often termed as an adaptation strategy (Black, Bennett, et al., 2011b) for avoiding vulnerabilities. Migration as an adaptation strategy is cited more in climate change and migration literature to explain migration as a good option to save the affected people from climatic events (Gemenne & Blocher, 2017; McLeman, 2014). In this sense, the CHT was a lucrative place for aspiring migrants because of its land and forest resources. Equally, Dhaka is also chosen by the migrant at having the work opportunities. The climate-affected people in the lowland and rural areas considered this as a place suitable for settlement, and the government and army intervention rendered the CHT more accessible and safer for Bengali migrants.

Another mediating factor that helped to institutionalize the migration is the social networks that link aspiring migrants with people who have already settled in the destination place (Curran, 2002; Massey, 1990; Massey et al., 1993). Climate migrants interviewed for this study outlined the help they received from their kin or from people of their local community in their migration to the cities and the CHT. Aspiring migrants activated connections to family, neighbours, and local villagers who had already settled in the cities and the CHT. In the case of the CHT, some Bengali people met with their neighbours and relatives and sought their help to settle in the region. As the Bengali population in the CHT has grown, the social networks have become more ubiquitous, enabling aspiring migrants to manage their own migration process. People can obtain more accurate information about the situation of the CHT through these extended social networks and imagine migrating and establishing their lives there. The communication and transport development is both a consequence and a cause of increased mobility of people to and from the CHT, and the ease of access makes it easier to settle in the area. The migration flows to the CHT are self-sustaining and no longer require an active government policy. The region is now also better connected with bus services to districts from where many migrants originated, including Noakhali, Feni, Chittagong, Comilla, Chandpur, and Barisal.

The conflict between the Shanti Bahini and the army has an indirect role in the Bengali migration to the CHT. Migration literature shows that conflicts and emergencies expedite the migration outflow because people move away to save their lives (Guadagno, 2016; Van Hear et al., 2018). But in the case of the CHT, the conflict situation influenced the Bangladesh policymakers to settle the Bengali people in the land of the tribal people in the hope that this would defuse the conflict. However, the Shanti Bahini turned their anger toward the Bengali settlers, burning their houses and sometimes injuring and killing them in attacks that were organized to frighten the settlers so that they would leave the CHT (Adnan, 2008; Ahsan & Chakma, 1989). There is little evidence to show whether Bengalis left the region due to the conflict; however, there are indications that Bengali settlers assisted others in their network to settle in the CHT in order to strengthen their security situation. In that sense, the long-standing conflict has played a role in fostering Bengali migration to the CHT.

Conclusion

It can be argued that several factors prepared the ground for the growth of migration to the urban areas and the CHT in the post-independence era. These composite factors are the intervening and mediating factors of migration analysis. Even though push factors such as floods, cyclones, coastal erosion, and drought have complicated the socioeconomic situation of millions of people in many parts of Bangladesh; without mediating factors the migration to the CHT would not have taken place the way it did. Because the CHT was not well connected to the rest of Bangladesh and the site of a conflict over self-determination, not many Bengalis would have contemplated moving here without assistance. Thus, the mediating factors such as government policy and assistance, protection by the army, social networks, and family connections worked to institutionalize the Bengali migration process. Most research has generalized the migration and settlement as state sponsored and planned settlement. There is no denying the fact that it was a planned and calculated device of the successive governments of Bangladesh; however, a systematic study of the mediating factors for the Bengali migration and settlement based on environment and migration literature has provided a new and alternative explanation which is almost missing in the domestic literature in Bangladesh.

Notes

1 The CHT as an excluded area means the distinct nature of the CHT, which should be only enjoyed by the tribal people. This was created by the British colonial government though the Regulation of 1900 Act.
2 The Amendment of the Rule 34 is related to the abolition of the special status as well as allowing the non-tribal people to settle, buy land and construct houses in the CHT.

References

Adnan, S. (2004). *Migration land alienation and ethnic conflict: Causes of poverty in the Chittagong Hill Tracts of Bangladesh.* Dhaka: Research & Advisory Services.

Adnan, S. (2007). Migration, discrimination and land alienation: Social and historical perspectives on the ethnic conflict in the Chittagong Hill Tracts of Bangladesh. *Contemporary Perspectives*, *1*(2), 1–28.

Adnan, S. (2008). Contestations regarding identity, nationalism and citizenship during the struggles of the indigenous peoples of the Chittagong Hill Tracts of Bangladesh. *International Review of Modern Sociology*, *34*(1), 27–45.

Adnan, S., & Dastidar, R. (2011). *Alienation of the lands of indigenous peoples: In the Chittagong Hill Tracts of Bangladesh*. Copenhagen: International Work Group for Indigenous Affairs. https://www.southasia.ox.ac.uk/sites/default/files/southasia/documents/media/oxford_university_csasp_-_work_in_progress_paper_15_adnan_dastidar_alienation_of.pdf

Ahsan, S. A., & Chakma, B. (1989). Problems of national integration in Bangladesh: The Chittagong Hill Tracts. *Asian Survey*, *29*(10), 959–970.

Black, R., Adger, W. N., Arnell, N. W., Dercon, S., Geddes, A., & Thomas, D. (2011a). The effect of environmental change on human migration. *Global Environmental Change*, *21*(1), S3–S11. https://doi.org/10.1016/j.gloenvcha.2011.10.001

Black, R., Bennett, S. R., Thomas, S. M., & Beddington, J. R. (2011b). Climate change: Migration as adaptation. *Nature*, *4780*(7370), 447–449.

Chakma, B. (2010). The post-colonial state and minorities: Ethnocide in the Chittagong Hill Tracts, Bangladesh. *Commonwealth & Comparative Politics*, *48*(3), 281–300. http://www.tandfonline.com/action/showCitFormats?doi=10.1080/14662043.2010.489746

Chatterjee, R. (1987). Cotton handloom manufactures of Bengal, 1870–1921. *Economic and Political Weekly*, *22*(25), 988–997.

Choudhury, G. W. (1969). *Constitutional development in Pakistan*. Vancouver: Publications Centre, University of British Columbia. https://archive.org/details/constitutionalde00chou/page/n5.

Curran, S. (2002). Migration, social capital, and the environment: Considering migrant selectivity and networks in relation to coastal ecosystems. *Population and Development Review*, *28*, 89–125.

Gemenne, F., & Blocher, J. (2017). How can migration serve adaptation to climate change? Challenges to fleshing out a policy ideal. *The Geographical Journal*, *183*(4), 336–347.

Goodhand, J. (2000). Research in conflict zones: Ethics and accountability. *Forced Migration Review*, *8*(4), 12–16.

Guadagno, L. (2016). *Integrating migrants in emergency preparedness, response and recovery in their host countries*. Training manual (Facilitator's Guide). International Organization for Migration (2016). Geneva: IOM. https://publications.iom.int/system/files/pdf/micic_reference_e-handbook.pdf

Guhathakurta, M. (2012). The Chittagong Hill Tracts (CHT) accord and after. In D. Pankhurst (Ed.), *Gendered peace: Women's struggles for post-war justice and reconciliation* (pp. 197–214). Hoboken: Taylor and Francis.

Höglund, K. (2011). Comparative field research in war-torn societies. In K. Höglund & M. Öberg (Eds.), *Understanding peace research: Methods and challenges* (pp. 114–129). Abingdon: Oxon.

Hugo, G. (2013). *Migration and climate change*. Cheltenham, UK: Edward Elgar Publishing.

International Organization for Migration (IOM). (2018). *World migration report 2018*. Geneva: IOM. https://publications.iom.int/system/files/pdf/wmr_2018_en.pdf

Islam, R., Schech, S., & Saikia, U. (2021). Climate change events in the Bengali migration to the Chittagong Hill Tracts (CHT) in Bangladesh. *Climate and Development*, *13*(5), 375–385.

Khandker, S. R., Khalily, M. B., & Samad, H. A. (2012). Seasonal migration to mitigate income seasonality: Evidence from Bangladesh. *Journal of Development Studies*, *48*(8), 1063–1083. https://doi.org/10.1080/00220388.2011.561325

Kuhn, R. (2003). Identities in motion: Social exchange networks and rural-urban migration in Bangladesh. *Contributions to Indian Sociology*, *37*(1–2), 311–337. https://doi.org/10.1177%2F006996670303700113

Lee, E. S. (1966). A theory of migration. *Demography*, *3*(1), 47–57.

Lee, S.-W. (1997). Not a one-time event: Environmental change, ethnic rivalry, and violent conflict in the Third World. *The Journal of Environment & Development*, *6*(4), 365–396.

Lee, S.-W. (2001). *Environment matters: Conflicts, refugees & international relations*. Seoul and Tokyo: World Human Development Institute Press.

Levene, M. (1999). The Chittagong Hill Tracts: A case study in the political economy of creeping genocide. *Third World Quarterly*, *20*(2), 339–369.

Marino, E., & Lazrus, H. (2015). Migration or forced displacement?: The complex choices of climate change and disaster migrants in Shishmaref, Alaska and Nanumea, Tuvalu. *Human Organization* *74*(4), 341–350. https://doi.org/10.17730/0018-7259-74.4.341

Martin, M., Billah, M., Siddiqui, T., Abrar, C., Black, R., & Kniveton, D. (2014). Climate-related migration in rural Bangladesh: A behavioural model. *Population and Environment*, *36*(1), 85–110. https://doi.org/10.1007/s11111-014-0207-2

Massey, D. S. (1990). Social structure, household strategies, and the cumulative causation of migration. *Population Index*, *56*(1), 3–26.

Massey, D. S., Arango, J., Hugo, G., Kouaouci, A., Pellegrino, A., & Taylor, J. E. (1993). Theories of international migration: A review and appraisal. *Population and Development Review*, *19*(3), 431–466.

McLeman, R. A. (2014). *Climate and human migration: Past experiences, future challenges*. New York: Cambridge University Press.

Mohsin, A. (1997). *The politics of nationalism: The case of the Chittagong Hill Tracts, Bangladesh*. Dhaka: University Press Limited (UPL).

Mohsin, A. (2003). *The Chittagong Hill Tracts, Bangladesh: On the difficult road to peace*. Boulder, CO: Lynne Rienner.

Mohsin, A., & Ahmed, I. (1996). Modernity, alienation and the environment: The experience of the hill people. *Journal of Asiatic Society of Bangladesh*, 41(2), 265–286.

Panday, P. K., & Jamil, I. (2009). Conflict in the Chittagong Hill Tracts of Bangladesh: An unimplemented accord and continued violence. *Asian Survey*, *49*(6), 1052–1070.

Reuveny, R. (2008). Ecomigration and violent conflict: Case studies and public policy implications. *Human Ecology*, *36*(1), 1–13. https://doi.org/10.1007/s10745-007-9142-5

Siraj, N., & Bal, E. (2017). 'Hunger has brought us into this jungle': Understanding mobility and immobility of Bengali immigrants in the Chittagong Hills of Bangladesh. *Social Identities*, *23*(4), 396–412.

Suhrke, A. (1997). Environmental degradation, migration, and the potential for violent conflict. In Gleditsch, N.P. (Eds.), *Conflict and the environment* (pp. 255–272). NATO ASI Series (Series 2: Environment), vol 33. Dordrecht: Springer.

Tripura, S. B. (2008). *Blaming jhum, denying jhumia: Challenges of indigenous peoples land rights in the Chittagong Hill Tracts (CHT) of Bangladesh: A Case study on Chakma and Tripura*. Masters Thesis, Universitetet i Tromsø, Norway. https://munin.uit.no/handle/10037/1535

Ullah, M. S., Shamsuddoha, M., & Shahjahan, M. (2014). The viability of the Chittagong Hill Tracts as a destination for climate-displaced communities in Bangladesh. In S. Leckie (Ed.), *Land solutions for climate displacement*, (pp. 215–247). London: Routledge.

Van Hear, N., Bakewell, O., & Long, K. (2018). Push-pull plus: Reconsidering the drivers of migration. *Journal of Ethnic and Migration Studies*, *44*(6), 927–944. https://doi.org/10.1080/1369183X.2017.1384135

Warner, K. (2010). Global environmental change and migration: Governance challenges. *Global Environmental Change*, *20*(3), 402–413. https://doi.org/10.1016/j.gloenvcha.2009.12.001

6 The migration and conflict interplay in Bangladesh

Introduction

Chapter 3 highlighted that Bangladesh is a conflict-prone country. Different forms of conflict from identity-based to political conflict engulf the country and endanger the peaceful situation. In urban areas, the nature of the conflict is civil strife, violence, and political turmoil. In the case of the CHT, the conflict originated due to the denial of the identity of the tribal people by the constitution of Bangladesh in 1972 (Mohsin, 1997, 2003). This conflict was termed ethno-political conflict and was between the Bangladesh army and Shanti Bahini for the right to self-determination of the tribal people in the CHT. Bengali settlement and migration escalated the conflict in the 1980s and 1990s and created ongoing social tension between the Bengali and tribal people. The migration of the Bengali people has established a new discourse of Bengali versus tribal in the CHT, which posits the two communities as polarized, divided, and conflicting. Based on multidisciplinary approaches to climate change–induced migration and conflict, this chapter explains how the migration of the Bengali people has generated social conflict in the CHT. Panday and Jamil (2009: 1066) described the conflict as "armed insurgencies between the tribal militants on one hand, and Bengali settler groups and the army on the other". This chapter does not refer to the current conflict as armed insurgencies, but rather labels it as a micro-level social conflict between and among the communities, for reasons discussed subsequently.

Climate change–induced migration caused significant changes in the urban areas and to the social, economic, and political system in the CHT. The most notable change is the demographic change due to migration, which has led to the population consisting of Bengali and tribal people in equal proportions. The tribal people see the equalization of Bengali and tribal people as a threat to their existence as they are losing power and control over social, economic, and political positions, as well as becoming marginalized day by day in their own land. The Bengali settlement has also contributed to resource scarcity, resource competition, and resource conflict. As migration flows gather pace, the relationship between tribal and Bengali people has increasingly become

DOI: 10.4324/9781003430629-6

affected by mistrust, misperception, hatred, and sometimes violence between and among the groups. The Peace Accord process nurtured hopes that ending the conflict between the Bangladesh army and Shanti Bahini would establish peace; however, the region continues to experience conflict between Bengali and tribal people over resources and social, economic, and political power (Panday & Jamil, 2009; Mohsin, 2003). The current conflict in the CHT is related to capturing and controlling the resources, control over political rights and social positions in public and private sectors and in every sphere of life. Despite the ongoing peacebuilding efforts (Chakma, 2017; Braithwaite & D'Costa, 2018), incidents of fear, insecurity, direct violence, killing, torture, and social division affect many locations in the region. While the outright war between the army and the CHT independence forces may have ended, micro-level social conflicts at the community level are affecting the peace and security of the people (Braithwaite & D'Costa, 2012). In this current form of conflict, the presence of the Bengali people is one of the key factors in contributing to conflict formation and escalation. The CHT has turned into an imposed multi-ethnic society where the Bengali and tribal communities have emerged as rivals engaged in conflicts for social position, political participation, and access to economic opportunities and services (Badiuzzaman, Cameron, & Murshed, 2011; Panday & Jamil, 2009).

This chapter engages with the argument that migration of people of a different background to an ethnically dominated areas causes mistrust, fear of losing livelihood options, and conflict (Reuveny, 2007). The demographic change due to the migration also causes resource scarcity, resource competition, and conflict (Kahl, 2006; Goldstone, 2002). The population pressure due to immigration generates environmental stress, resource scarcity, and resource capture in the host society, and depriving some groups from access to resources fuels group competition (Homer-Dixon, 2010). Sometimes the migration process generates a heightened feeling of ethnicity and of "we" and "they" between the migrants and local people, which then becomes a source of conflict in a host place (Reuveny, 2008). Besides environmental scarcity, demographic change also has the potential to generate social experiences of exclusion and a discriminatory political and economic system (Kahl, 2006). In the case of the CHT, the migration of people with a different cultural background to an ethnically distinctive area has not only put pressure on existing resources but also complicated social, cultural, and political relationships.

This chapter, thus, focuses on the relationship between migration and micro-level social conflict in urban areas and the CHT, whereby micro-level social conflict can extend to communal violence, attacks, killings, burning, rape, and human rights violations. In the case of the CHT, this chapter argues that migration and settlement of the Bengali people have led to land grabbing by the Bengali settlers, civil and military elites, and other influential people. The new migrants have captured resources and competed with the tribal people for social and political positions in the CHT. The land grabbing and resource capture consequently contributed to new forms of social relationships characterized by widespread mistrust and polarization, and sometimes violence.

This chapter first presents the process of conflict occurring in the urban centers for the increasing migration. This conflict is not violent in nature but manifested violently sometimes. This chapter then explores the migration and conflict interplay in the CHT in Bangladesh where conflict has been caused significantly by the Bengali migration and settlement.

Migration and conflict interplay in urban areas

Population pressures in urban areas and suburban areas in Dhaka are mounting with the development that has been achieved through industrialization in the country. From only 200,000 people in the year 1947, it has risen up to nearly 11 million in 2001 (Hossain, 2008). The population is further projected to be around 27 million in 2030, and this rapid growth compared to even the other urban areas in the country is causing some apparent problems with regards to housing and employment, among other issues (Khaleda, Mowla, & Murayama, 2017). This migration from the climate-affected areas to Dhaka triggers existing conflict and generates new forms of conflict in the urban center. This section describes how migration and conflict interplay in urban areas.

Compromised residence and basic services

The rapidly growing rural migrant populations in the city have also contributed significantly to this rising population problem, and the resulting housing problems that they suffer are only one of many issues faced by these comparatively less prepared residents of the city, as opposed to people who have their lives properly settled in the city. These migrants lack first and foremost formal tenure rights and are therefore often subject to eviction by the government and corporations. This is even more so when the pressure mounts from rapid development and industrialization in the city (Ahmed & Meenar, 2018). The size of this population, even before 2015, was nearly a third of the total population of 12 million in Dhaka, and they have since taken residence in around 4,500 slums and squatter settlements (Paul, 2006; Paul & Ramekar, 2018), which largely remain unauthorized and thus prone to eviction maneuvers (Islam & wa Mungai, 2016). Findings reveal that the residency of these slum dwellers is seen as secondary to other issues in their area by policymakers; as indicated by Lata,

> the local and central government officials categorize slum dwellers as encroachers and criminals, who pose a direct threat to an orderly, clean and green city. Hence, they cannot be allowed to exist in the city. Additionally, the state has shifted the development of land and housing

> markets to real estate developers, following a neoliberal economic model. Consequently, a few powerful developers control Dhaka's land and housing markets, only supplying housing for the growing middle class. Access to these houses is far beyond poor people's reach.
>
> (Lata, 2020)

This further indicates that the housing rights of these impoverished urban people are denied both by the state and by the market in Dhaka city. Lack of basic rights such as water, sanitation, and electricity becomes a source of conflict. Key informant 15 mentioned that "slum people suffer from serious water scarcity. They are to spend money for getting access to water. Sometimes, they protest against the local government for getting water" (key informant 15). In fact, the slum people illegally live and construct their houses; thus, they do not get access to basic services, such as water. A study conducted in slums argued that the scarcity and increasing demand of water by the climate migrants in Dhaka city triggers and aggravates conflicts among different actors in the slum's areas of Bangladesh (Nurul & Mohammad, 2014). An expert working in the development field argued that most of the slum people are involved in the conflict (verbal abuse, small-scale fights, and harassment). The scarcity of resources, poor governance, and lack of political influence of the slum people force these sections of people to lead such a conflictual everyday life (Expert 7).

Poverty and conflict nexus

The migration of these rural populations in search of better opportunities and livelihoods leads to a mismatch of skill sets and available scope for income if the resulting poor environments and unsanitary settlements that they end up taking residences in are any indication (Khan, Kundu, & Yeasmin 2021). Their generally low skillset stemming from their significantly low literacy rates (Latif, Irin, & Ferdaus, 2016) can lead to problems besides a lack of decent scope for income generation. A climate change and migration expert outlines that

> migrant people lack access to the land needed for permanent residence, nor do they have the stable income needed for renting apartments, and thus they reside in public land or land which is unoccupied for the time being till development initiatives (i.e., public or private) force them to move elsewhere. This situation generates conflicts and violence.
>
> (Expert 9)

Poverty and poor living conditions push many of the slum dwellers to become involved in violent activities for accessing basic services. They also become victims of human rights violations by local Mastans and law enforcement agencies. Expert 8 narrated that,

> Slum dwellers are displaced people from their original place of land. They live in such a place to save their family and feed their children. But they can't live in peace. Local Mastans and police harass them frequently. Some of the slum dwellers are involved in some sort of illegal business. But majority of them are victims for this. Sometimes, local Mastans use the slum people to run their illegal businesses. Slum becomes their safe place to sell the drugs and other illegal items.
>
> (Expert 8)

The access to these facilities does not always translate to better living conditions and affordability. Low literacy rates lead to poor healthcare and sanitation decisions, and residents of slums are as a result prone to common diseases that result from unsanitary lifestyles and living spaces (Rahman et al., 2021). The unstable nature of the income that the people in these households have means it experiences decline at intervals. The resulting decline in purchasing power for these slum-dwelling households result in serious compromises to the health of these people living in unsanitary and cramped conditions of urban slums, especially during the Covid-19 pandemic (Koly et al., 2021).

Survival, extortion, and conflict

The people have attempted to look for legal avenues for income-generating opportunities via low-skilled labour wage and other work with low-skill requirements. That being said, the majority of slum dwellers do not have the necessary access to formal economic opportunities provided by public or private organizations, save for a few exceptions who have been able to engage in the garment sectors and other low-wage work (Lata, 2017). The majority of the studies conducted on this topic within Dhaka indicate that policy frameworks thus far have not sought to solve the problems with regards to the housing and employment opportunities for the poor, urban slum-dwelling populations. It reflects that migrant slum dwellers struggle to survive in the new place. Moreover, Mastans and police often demand extortion. The majority of the key informants mentioned extortion (i.e., *chanda*) to survive in the slums. This frequent demand of extortion generates conflicts. We investigated the matter further in a bid to understand the political economy of survival in the slums. Another expert argued that,

> Along with other bills, the slum dwellers are to pay any money to the Mastans and sometimes to the police to live peacefully in the slums. Without regular extortion, the Mastans come and threaten them. Sometimes, the Mastans put excessive pressure. They also pay extortion to police and Mastans if they like to start a small business.
>
> (Expert 15)

In the slums, Mastans are the agents of political wings who are actively involved in politics. These influential groups seek money from the slum dwellers on a regular basis. At the same time, they sometimes force the slum dwellers to attend political programs such as Hartal, political processions, and agitations. As the slum dwellers are powerless, they become victims to participate in violent political processes due to outside pressure. An expert, a history professor from the University of Dhaka, argued that,

> some sections of slum dwellers are highly involved in violent political processes. They are easily picked by the political parties to become involved in violent activities. They are also involved in this violent process by getting cash money from the political leaders.
>
> (Expert 10)

The often violent nature of the overall political system of Bangladesh thus often involves political clients and active supporters attending in every political process to vandalize the public property for attaining political goals.

Resource scarcity and conflict nexus in urban areas

The urban areas, particularly Dhaka, are expanding rapidly. Tides of migrant people from the climate affected areas of Bangladesh further push the rising population and existing resources. The unrealistic planning has also been adding serious strain on the resources and generating vulnerabilities in the cities. The pressure from the migrants and the resulting resource scarcity often trigger the conflict and violence in the cities. Existing literature argues that resource scarcity and population pressure generate conflict and violence in a society where the governance system fails to make balance between the rising population and resources (Homer-Dixon, 2010; Kahl, 2018). This unmanaged and uncontrolled population pressure of Dhaka is now creating mounting pressure on the city's resources, which is consequently affecting the people's already decreasing living standards. Resource scarcity and population pressure often cause civil strife, unrest, and violence in the city. Different studies outline that poor people in the slum are involved in some activities that generate violence and conflict (Saha, 2012). While asking this issue to an expert, an independent researcher in the field of environment and security argued that,

> The city's natural resources and human environment are under serious threats and vulnerabilities. The governance is responsible for these threats and vulnerabilities. As the governance system is poor and incapable of providing basic services to all city dwellers, a sense of deprivation always exists among the deprived section of people. This deprivation and acute poverty push some of the slum people to become involved in conflict and violence.
>
> (Expert 7)

A key informant also explained the issue of resource scarcity and conflict connection in the slum in Dhaka city. While asking the question of their involvement and experience of conflict in the slums, the key informant highlighted that

> people in slums are powerless and mostly deprived of getting secure jobs. In many cases, people are involved in non-formal and small jobs which are not secure. Thus, they face insecurity in managing livelihoods. Some people are also victims of torture and human rights violations. In such a case, some of them get involve in outlaw activities. But this is not applicable for all people live in slums.
>
> (Expert 13)

Indeed, migration and conflict in the urban city due to the pressure of climate change-induced impacts is a complex issue. Displaced and migrant people do not directly cause conflict and violence, but they cause some triggering factors in the cities that generate conflict and violence.

Lack of political presence, and use as the actors of violence

Displaced and migrant people in the urban areas are powerless and hardly get scope to participate in political activities. However, the slum dwellers are often used as the toys of the influential political leaders and Mastan. Because the land they generally occupy for their residence is largely under the ownership of elites and agencies, this often leads to a lack of a stable residence for these people. The political participation of these people become very limited when they are under frequent pressures rising from low-wage income and land acquisition and resulting evictions (Nahiduzzaman, 2006). Moreover, their housings are generally tied to their negotiations with influential people who own the land. The pressure these slum dwellers receive from a one-sided negotiation with their de facto landlords can limit their political participation in the governance system even further than it already is (Hossain, 2013). Coalitions of these populations resulting from shared experiences have taken measures to establish some form of political grounds for achieving their collective demands through political channels. The Coalition for the Urban Poor's (CUP) Basti Basheer Odhikar Surakha Committee (BOSC) is a notable example in this regard. These initiatives, however, have not been able to move beyond merely connecting with a local ward commissioner to pitch their demands, and as such their political reach still remains largely limited based on their relationship and ties with these local representatives (Banks, 2008). Discussion with the conflict experts unveils the issue of involvement of the slum dwellers in conflict and violence, which is merely happening for their powerlessness and lack of legitimacy to live in the slums. An expert, a professor of international relations, argued that,

> The slum dwellers are migrant people who come from other parts of Bangladesh searching for their shelter and livelihood in slums. As these places are not registered, they are not entitled to receive basic services, on top of that they always live in fear of eviction. In some cases, they arrange their houses by giving money on a monthly basis to Mastans and influential people who generally control the entire management of slums. This informal system of settlement by the migrants is more prone to face conflicts and human rights violations.
>
> (Expert 12)

Indeed, internal migration of the displaced people from the rural areas to urban areas acts as a source of conflict as and when they put pressure on the existing resources, and they are deprived from receiving basic needs. As displaced people also do not have access to political power, Mastans and police sometimes harass people in the slums. In such a situation, they adopt every possible option to survive in the slums. When asked about their involvement in conflict and outlaw activities, key informant 20 argued that "they are powerless and helpless people. Thus, they are often influenced by the Mastans, political cadres and police to do some outlaw activities. They confess their helpless situation." The city governance system hardly provides its services to these people.

Criminal involvement

The involvement of slum dwellers in crimes and violent conflicts can be seen as a detrimental side of such developments in the city. A sociologist from the University of Dhaka argued that

> With high criminality being observed in families occupying the urban slums in Dhaka, children are the most affected in the long term. The economic issues in these families lead to a lack of proper parental upbringing of the children within these families, and this results in the children becoming involved in crime.
>
> (Expert 2)

The involvement of children of these slum areas in crime and drug abuse is at an alarming rate of over 75 percent as of 2015, where the children are becoming involved in drug abuse, hijacking, and theft (Kamruzzaman & Hakim, 2015). This involvement in less than lawful activities extends to the slum-dwelling population of the city in general. The local political leaders in the city have used the opportunity created by the lack of power of these populations for their own political interests. The previously mentioned lacking of these populations of political capabilities does not result in a lack of political awareness of these slum dwellers, since over 50 percent of them are politically conscious of the power dynamics in local politics. Their direct participation in various activities like meetings of political parties and picketing during political strikes demonstrates how their roles in local and national politics have developed based on their set of available opportunities (Hasam et al., 2017). However, the slum dwellers have an alternative perception about the involvement in criminal activities. A key informant in a slum argued that "they are the victims for their involvement in criminal activities. They are poor as well as highly deprived of accessing their basic necessities. Influential people and Mastans use them to conduct outlaw activities" (Expert 10).

Discussion on migration and conflict in slums

It is to note that the migrants arrive from rural areas in search of better income opportunities but are often locked into a cycle of economic decline and/or stagnation. Their involvement in often unlawful activities can naturally be seen as a result of their low legal avenues for income and the helplessness that occurs as a result. This can be seen via the lens of the human needs theory that Burton has claimed (Christie, 2009), as people seek opportunities for better lives and to meet their basic needs based on available avenues, with legal implication being sidestepped in favor of basic survival. The conflict potential generated by these migrants, however, cannot be ignored; as Saha has indicated,

> additional factors which exacerbate violent conflicts in the slum areas of Dhaka. These factors are: a lack of basic necessities, deteriorating living conditions, the prevalence of crime and the availability of small arms, and a lack of justice.
>
> (Saha 2012)

The preceding findings indicate that the conflicts occurring within the slum-dwelling population stems from their predicament and conditions previously mentioned above. The availability of small arms is a concerning find considering the extent of involvement of these people in unlawful activities. The lack of justice can be seen as stemming from their lack of strong political representation in governance and decision making when the distribution of public benefits are concerned. The families of these slum dwellers also suffer from familial relationship issues, as records of domestic violence in families of urban slums have been found to be substantial and nearly over 50 percent (Chowdhury et al., 2018). These high rates of domestic violence in these families in the form of intimate partner violence (IPV; that is, violence between spouses) have resulted in behavioral problems in their children, which can to a reasonable degree explain their deviant behavior in their involvement in unlawful activities and gang violence (Kalim et al., 2017). Such findings reveal the conflict potential of this population in Dhaka. If their conditions persist, unwanted developments in resource availability may push them toward more drastic measures in the near future. These migrant urban slum populations are sharing finite resources with other residents of the city. If the avenues call for unlawful activities with much higher payoffs than what these populations are currently living on, situations may well deteriorate with manifestations of violent conflicts.

Migration and conflict interplay in the CHT

The conflict in the CHT has been viewed as an ethnopolitical conflict which originated due to the denial of the recognition of the right to self-determination. Over time, the nature of the conflict has changed due to the migration of the Bengali people as well as the changing sociopolitical environment in the region. The following section describes the interplay of migration and conflict in the CHT.

Resource sharing and conflict

The migration and settlement of the Bengali people to the CHT has impacted resources, resource sharing, and conflict. The interview with experts provided information about resources and land that offer some insights into the underlying causes of social conflict and violence. In a multi-ethnic society, resource scarcity may generate conflict and violence when groups compete for resources and seek to capture resources by depriving others (Homer-Dixon, 1999). Reuveny (2007: 659) argues that "the arrival of environmental migrants can burden the economic and resource base of the receiving area, promoting native-emigrant contest over resources." In the CHT, Bengali migration and settlement has led to resource scarcity and the situation of ethnic conflict and violence. The key informant interviewees were asked about the forest resources and land in the CHT. The experts were asked about livelihood opportunities in the CHT.

Key informants belong to the Bengali community living in the CHT perceive that there are sufficient land and forest resources, and hence opportunities for earning a living in the CHT. Indeed, the migrant Bengali people in the CHT are involved in the agricultural sectors in the CHT along with small businesses and service sector employment. One of the key informants mentioned that "Bengali migrants turned their profession towards business and service sectors, which is helping them hold a strong position in the region" (Expert 7). In contrast, the majority of the key informants belonging to the tribal community perceive that there is not enough land in the CHT and forest resources in the CHT. A key informant from the tribal community outlined that "Bengali migrations are contributing to reducing the percentage of land and forest resources. Moreover, the policy of the government and private organisations of setting up business infrastructure are also contributing to the scarcity of land and forest resources" (Expert 10).

Land and forest resources are the major sources of livelihood of many tribal people. Historically, tribal people are dependent on forest land for *jhum* cultivation, which forms the basis of their livelihood (Roy, 2002; Thapa & Rasul, 2006). But the per capita agricultural land and forest resources have diminished in the CHT due to the competition for these resources from Bengali settlers and elite people. This shrinking of per capita land has been possible as Bengali settlers have occupied land for their houses and cultivation. The land scarcity has made many tribal people insecure and frustrated (Adnan, 2004, 2008). This theme also emerges from the interviews with key informants, where one academic CHT expert describes the "dynamics" of resource scarcity as caused by the process of settling the Bengali people in the CHT, allocating them land, and putting pressure on the existing resources (Expert 3). According to the expert, "tribal people who depend on agriculture and forest are the worst victims of resource scarcity in the CHT. They are most likely to be deprived of their basic needs, such as food, house, and health facilities" (Expert 3).

This is confirmed in another interview by the director of a research organization who made the following point:

> The CHT now suffers from serious resource scarcity. The per capita land is being decreased. The jhum cultivation, which is the major source of livelihood for the poor tribal people, is interrupted by the population pressure and excessive pressure on the hill. Nowadays, some tribal people are equally responsible for the exploitation of the resources of the CHT. The worst victim of this resource scarcity is the marginal tribal communities who only survive by the jhum cultivation and forest resources.
>
> (Expert 12)

Later in the interview, the informant connected the shrinking resources in the CHT with the intervention of different stakeholders, including settlers, powerful elites of both Bengali and tribal backgrounds, and multinational corporations.

Land grabbing and conflict

The dispute over land ownership and land grabbing is at the heart of most of the conflicts in the CHT in the post Peace Accord period (Adnan & Dastidar, 2011; Roy, 2000; Amnesty International, 2013). The notion of land grabbing refers to systematic as well as forceful capture of land (Feldman & Geisler, 2012). In the CHT, government, elites, corporations, and local people systematically seize land designated as *Khas* (government-owned fallow land without individual property rights), land for conservation, rubber plantations, and forest reserves (Adnan &d Dastidar, 2011; Halim & Chowdhury, 2015). The land is acquired in a nontransparent way that violates the rights of the people who depend on it (Adnan & Dastidar, 2011). The key informants were asked about land grabbing in the CHT. Interestingly, key informants from both the communities shared similar information about land grabbing in the CHT.

Land grabbing commenced during the insurgent movement (1976–1997) when successive Bangladeshi governments granted land to settlers under a formal titling process called *Kabuliyat*[1] (Adnan, 2004). Bengali settlers used the allocated land to build a dwelling and for agricultural purposes. In some instances, government officials deliberately and systematically violated the existing principles of land ownership in the CHT and grabbed land for their personal property (Adnan, 2004; Roy, 2000). On many occasions, the state acquired land for development purposes and declared portions of forest land "reserved" for conservation and protection for the environment. Reserving forest land has the effect of preventing tribal people from entering the forest for resources extraction. This process also leads to the displacement of the tribal people from their homes and from cultivable land (Adnan & Dastidar, 2011). The local governments in the CHT sometimes declare land as *khas* land and allocate it to Bengali settlers. For example, one study outlines that "the government has taken more than 14,000 acres of forest land in the Bandarban districts and allocated it to the Bengali entrepreneurs" (Ullah, Shamsuddoha, & Shahjahan, 2014: 205). There are many other instances of taking land from tribal owners. Barkat (2016: 132–4) argues that land grabbing equates to the land alienation of the tribal people in the CHT. Various techniques have been used in the process of land alienation including fake documents, forceful eviction, marking the land as *khas*, and developing eco-parks, national parks, or social afforestation projects.

Land grabbing happens on various levels and involves both ordinary people, sometimes unwittingly, and powerful private or public sector actors. One key informant commented on land grabbing that "the Bengali settlers, private companies, army, smugglers above all the 'conservation' policy of the government combine in grabbing land in the CHT" (Expert 2). Companies and powerful elites sometimes manipulate the existing law of land allocation with money and power. They seize land in the name of development to expand tourism and agribusiness. Local people – both tribal and Bengali– are marginalized and increasingly squeezed out of land ownership. The Human Rights Report (2012) mentions that

> Destiny Group illegally encroached on some 5500 acres of land in Lama Upazila in the Bandarban district for their plantation project. Ignoring the Bandarban Hill District Council Act and other laws related to land administration, they planted some 1.5 crores [15 million] trees in the area.
>
> (Barman 2013: 100)

The Human Rights Report (2016) mentions that in 2016 over 15,000 acres of lands belonging to indigenous peoples were under the process of acquisition, mostly for the establishment of special economic zones, special tourist zones, and forest reserves (Chowdhury & Chakma, 2016: 28).

This form of land grabbing is planned and based on force and power. Another key informant argued that "the companies buy the consent of police and administration with bribes and then erect a signboard of ownership on the land to announce that no one is to enter the property without permission" (Key informant 13). They employ local Mastans (outlaw people) to threaten people who oppose the company's land grab (Mohsin & Hossain, 2015: 100). A key informant mentioned an interesting point about the process of land grabbing. She argued that "In the process of the resource scarcity, the elites and political authority sometimes change the property rights issuing fake documents and depriving the ethnic minority people and make it their own" (Expert 4). This system of deprivation may cause the expulsion of the minority people and consequent ethnic violence (Homer-Dixon, 1994: 10). In the CHT, the local people, particularly tribal people, depend on the land and refuse to sacrifice their land rights. Some tribal people move to the deep jungle hoping to get access to the forest resources for their livelihood (Expert 14).

Land grabbing is also complicating the sharing of common resources, such as catching fish in the lake and utilizing the forest resources. An expert argued that

> The private companies and influential people prevent other people from entering their recently acquired land to collect the resources for their livelihood, however this land has been used as common property for their daily livelihood by the tribal people for decades or more.
>
> (Expert 2)

This is the customary law and practices of land rights in the CHT (Adnan, 2004: 36), which emphasize sharing and caring for the land, and maintaining it in good health for the use of future generations. Private ownership of land is a foreign concept for many tribal people. Land grabbing by the Bengali settlers and civil and military bureaucrats is also weakening the position, identity, and cultural practices of a tribal community by detaching them from land and resources. Thus, grabbing land is threatening the concept of "common property"[2] and weakening the culture of the tribal people. Some researchers have referred to this entanglement of the cultural and livelihood system as "cultural annihilation," "ethnocide," or "genocide" of the tribal people in their land (Chakma, 2010a; Levene, 1999).

Both Bengali and tribal key informants perceived that land grabbing is generating conflict in the CHT. Almost all the tribal informants saw land grabbing as a source of conflict, as did almost two-thirds of Bengali respondents. Conflict formation is due to land grabbing as an expression of unequal competition for scarce land between groups, which is enabled by poor governance including fabrication of land title documents. An informant also viewed "land grabbing as an ongoing problem and a source of conflict in the CHT" (Expert 7). The process of generating conflict is complex. Asked how land grabbing creates conflict, a director of a rights-based organization in Bangladesh explained:

> The settlers have illegally occupied land in addition to their standard allocation of 3–5 acres of land by the government. Sometimes, Bengali settlers create a panic situation by burning the house, property, and forest and physically harming some tribal people who oppose land grabbing.
>
> (Expert 6)

Tribal people have limited choices beyond fighting or retreating into the more isolated areas of the forest. An informant mentioned that "Those who attempt to sell their property to outsiders find it difficult to prove their ownership of the land" (Expert 8). This information has been further explained by an expert, a history professor at the University of Dhaka:

> During the process of selling land, the Bengali people ask the tribal people to show their legal papers. The tribal people in many instances failed to provide any papers. As a result, the Bengali people bribe the people of the land office to prepare the paper using their name. Eventually, the Bengali settlers become the new owners of the land.
>
> (Expert 2)

Civil and military bureaucrats are involved in preparing fake documents for *khas* land and making it possible for powerful individuals and organizations to acquire land that is categorized as protected (Roy, 2000). Law enforcing agencies often fail to play a constructive role in managing the conflicts over land. Chowdhury (2012) identifies three ways the government is depriving the tribal people of their land rights and generating land disputes: "(i) non-recognition of customary resource rights and community ownership, (ii) introduction of private ownership based on title deeds as opposed to oral tradition; and iii) illegal settlement and grabbing of the Jumma land by government authorities" (Chowdhury, 2012: 39). Adding to the pressure on land in the post Peace Accord era is the return of tribal people who had left during the conflict between the army and Shanti Bahini to the CHT (Amnesty International, 2013). The tribal political leaders demanded that the land be returned to the refugees by the Bangladesh authority which gave it to Bengali settlers (Amnesty International, 2013; Adnan & Dastidar, 2011). The government of Bangladesh has helped to repatriate the refugees, but in many cases failed to return their land and home.

Overall, it is argued that migration and population growth in the CHT have increased pressure on the land, and land is becoming a scarce resource and a source of conflict, particularly in the eyes of tribal research participants.

Attempts to defend land rights, traditional or newly acquired, bring tribal and Bengali residents into conflict. Adding to this are powerful economic and political interest groups taking advantage of poor governance practices and land ownership legislation that is biased against customary ownership. Land grabbing and the growing scarcity of land have caused conflict over the ownership, use, and possession of the land that continues in the post Peace Accord period.

Social and political conflict

The CHT has been marked by conflict between Bengali and tribal residents and their respective political parties, including the government and the army. One consequence of this is a growing social polarization in CHT society, where mistrust, religious extremism, discrimination, and marginalization of tribal people take place (Choudhury, Islam, & Alam, 2017). Although discrimination, marginalization, and polarization are separate terms used in the literature, they are interconnected and generated due to the policy of the government and social practices of the people in the CHT. The following section explores how the study participants view the social implications of the conflict.

Social and political polarization

Polarization results from the "interaction of within-group identity and across-group alienation" (Esteban & Schneider, 2008: 132). It generates social and ideological separateness among and between groups. In the CHT, Bengali settlers and tribal people are highly polarized. In some areas the Bengali and tribal people live side by side, but they do not maintain social interaction or good relationships, which is one of the pressing problems in the post conflict social relationship in the CHT. It is well understood that a lack of interaction between communities encourages conflict and rift in the society (Ramsbotham, 2005). In the CHT, the low level of social interaction between the communities is resulting in communal violence and conflict (Choudhury et al., 2017). The past conflict between the two communities and land grabbing by the Bengali people are responsible for the low level of interaction, which is producing polarization in the society. The conflict over land and the broader political conflict over ownership of the CHT have affected social interactions and community relationships.

The situation of community relationships

Community relationship is understood by "trust," "preference as neighbour," and "relations with people with different religion." In a social system, trust between communities and relationships with other community people is the basis for harmonious relationships, social security, and the presence of peace (Jeong, 2005). Bianco (2017: 1) referred to this process as "building cultural dialogue" in a multiethnic society for sustainable peace. Key informants in both the communities mentioned that their intracommunal trust and relationships are strong, but intercommunal trust and relationships between the Bengali and tribal people are weak. Due to the lack of trust, both the communities prefer to have their own group of people as neighbours. A key informant argued that "tribal respondents do not like to live close to Bengalis, whereas the Bengali people trust and like tribal people as neighbours" (Expert 2). Another key informant from the tribal community argued that "This current relationship pattern is the outcome of the long-standing conflict, grabbing land and resources from the tribal people, and a series of small-scale communal conflicts between Bengali settlers and tribal groups" (Expert 13). The conflict situation has also acquired a religious dimension as places of worship, such as Buddhist temples, have been destroyed on a number of occasions in the CHT. However, the Bengali key informants mentioned that they have a good relationship with people with other religions. A Bengali key informant mentioned that "the tribal people also maintain a low level of relationship with people with different religions" (Expert 4). This indicates that the two groups of respondents in the CHT see the relationship with the other group in very different ways.

Interviews with experts provide further insight into how and why the CHT society has been polarized over time through government intervention. One expert suggested that "polarisation is the direct result of the government decision to migrate and settle Bengalis" (Expert 3). The Bengali migration was a conscious effort to make the CHT a multiethnic society which has brought many changes, such as the introduction of the Bengali language in educational institutions, textbooks containing the teachings of the Islamic culture, rapidly growing tourism, and new institutions – activities which are perceived by some tribal people as undermining tribal culture (Barkat et al., 2009; Uddin, 2010). Hence, these developments, which are presented by the government as progress and modernization, face significant opposition from tribal people. One expert, a director of a rights-based organization, explained the polarization between Bengali and tribal people as a consequence of "the Bengali settler increasingly getting a dominant position in the economic, social and political context" (Expert 6). This was observed during fieldwork in the 10 study locations in the CHT. Bengalis visibly dominate commercial centers in the towns as they have access to capital and connections with major cities in Bangladesh.

Different levels of mistrust and willingness to engage between the Bengali and tribal communities in the CHT might be explained by their different experience of the conflict in the area. In the post Peace Accord period both communities experience conflict and violence. For example, Choudhury and Hussain (2017: 130) mention "more than 500 cases of violence have occurred between the Bengali settlers and tribal people during 1997–2014." Tribal respondents in this study have been affected by violence more than Bengali respondents. Polarization and social division are clearly a consequence of violent conflict in the CHT, but also act as a force to create communal conflict in multiethnic societies (Esteban & Schneider, 2008; Forsberg, 2008). In analyzing the dynamics of the CHT conflict after the Peace Accord, Choudhury and Hussain (2017: 128) found that "the conflict issues are often small and latent. Due to the past conflict memories and polarisation, these small incidents turn into a violent conflict."

Islamization

Another consequence of the demographic transformation of the CHT is the Islamization[3] of the region, which is promoted by the Bengali Muslim population. Islamization is meant to increase the number of Islamic institutions such as mosques and madrassas and increase sponsoring of Islamic cultural practices in the CHT. The *Oxford English Dictionary* defines Islamization as "both a cultural and religious phenomenon. It contains both the conversion to Islam and expansion of the Muslim culture" (Peacock, 2017). While collecting field data in the CHT, I observed a small group of people wearing traditional Islamic dress appeal to a passerby for a donation to build a mosque for their community. Sometimes they support their request by reciting verses from the Holy Quran. Noticing this practice in several locations, I asked my research assistants about the issue. They explained that it is now a common practice to build mosques and arrange religious education events (*Waz-Mahfil*[4]) in the CHT.

In interviews, several key informants expressed concern about the Islamization of the CHT. One informant, a political leader of tribal background, suggested that it was part and parcel of the Bangladesh state's design to control the region and its population:

> As the bourgeois government sent the Muslim people to the CHT, the main task of the Muslim people is to promote the religion. In this sense, fundamentalism is increasing in the CHT. The tribal people are scared of this religious expansionism in the CHT.
>
> (Expert 5)

An expert argued that Bengali settlement and Islamization go hand in hand in the CHT. He suggested that only Muslim people were encouraged and supported to settle in the CHT, and this proved that the settlement scheme was "a device to spread Islam and dominate the tribal people in their own land" (Expert 5).

The process of Islamization in the CHT was started in the Pakistan period and accelerated in the 1980s with direct assistance from Saudi Arabia and through the formation of the nongovernment organization *Al-Rabita* in the CHT. It is alleged that M. A. Kashem (an antiliberation war criminal in Bangladesh and leader of the political party Jamat-I-Islam) initiated this

NGO to convert the tribal people to Islam (IWGIA, 2012). According to IWGIA, the organization provides a large amount of money for the enhancement of Islamic cultural practices in the CHT. The number of Islamic institutions in the CHT grew from 40 mosques and two madrassas in 1961 to 592 mosques and 35 madrassas in 1982 (Mohsin, 1997: 179). Since then, the establishment of Islamic institutions has increased rapidly, and recent data show that there are around 2,297 mosques and 1,552 madrasas (IWGIA, 2012: 18). The construction of mosques and madrassas requires land and adds to land disputes. In some instances, religious practices and rituals are used by their followers to antagonize tribal people. It is also alleged that some Muslim young people lure the tribal girls away from their families to marry them. Chakma and D'Costa (2013) argue that this constitutes a practice of "forced marriage" in order to convert tribal girls to Islam in the CHT (D'Costa, 2014: 28). All these social and religious practices between the tribal and Bengali people are generating deeper social polarization and conflict in the CHT.

Discrimination

In the literature on ethnic conflict, it is well recognised that discriminatory policies of the government toward any community, deprivation, exclusion, and systematic extinction processes cause serious civil strife, unrest, violence, and conflict between the government force and ethnic groups, and/or between and among the communities (Gurr, 1993, 2000; Fisher, 2016; Taras & Ganguly, 2015). Discrimination may exist in a society in which multiple ethnic groups live with division and polarization, and the state fails to ensure justice among the communities (Gurr, 1995; Gurr & Moore, 1997). The level of discrimination or discriminatory behaviour also reflects the level of conflict in society (Northrup, 1989; Tajfel & Turner, 1979).

The key informants from the Bengali and tribal respondents perceived that discrimination exists in the CHT, with a slightly higher level of agreement with the statement among Bengali respondents. An informant from the tribal community mentioned that "they were dominating in the CHT. But after the Bengali migration, they have been depriving and placing into the disadvantaged position in accessing to services in the CHT" (Expert 15).

In the post Peace Accord era, the issue of discrimination has risen to the top of the political agenda and public debate as Bengali and tribal people are almost equal in number. Thus, any visible policy of the government that favours one group, such as ration allocations to the Bengali settlers, generates a sense of discrimination among tribal people, particularly if they lead a miserable life. It was pointed out by one expert, a political leader with a tribal background, that there are many poor people in the tribal communities who never receive food rations from the government (Expert 5). Expert 10, a chairman of a research think-tank, explained this issue of discrimination as a consequence of systematic privilege given to the Bengali settlers (Expert 10). Expert 7, a director of an NGO of indigenous background, contributed further detail to understand what the systemic privilege is about:

> The most discriminatory policy of the government is the deployment of a vast number of soldiers in the CHT to protect the Bengali people. If the government withdrew the army from the CHT, I am sure that the Bengali people would not be able to live there. They will return to their place of origin.
>
> (Expert 7)

Discrimination exists at all levels of the political system including participation in the local administration and securing positions in the political parties in the CHT. Expert 6 suggested that "in future, the tribal people may not be represented as elected representatives at the local level through to the national level as a result of the number of the Bengali settlers increasing. The tribal contestants in many areas will not win elections in the future if citizens vote along ethnic lines" (Expert 6). In fact, this process of exclusion from political participation is not an example of discrimination because no one is actively preventing tribal people from contesting elections and holding political positions at the local and national level. But the tribal people may consider that they might lose all scope in the future as they are becoming a minority and economically weak in the CHT. I observed while undertaking the field study that Bengali people dominate in the business sectors due to their financial strength and the administrative support from the local government.

Yet the Bengali key informants showed a palpable sense of being discriminated against. This is exploited, or perhaps fostered, by new Muslim Bengali political movements in the CHT, such as *Somo Adhikar Andolan Parishad*[5] (Equal Rights Movement Party) and its student wing, *Parbatya Bangali Chatra Parishad*[6] (Hill Bengali Students' Council). These groups oppose the CHT Accord, which they see as infringing on the rights of Bengali people in the region. Their efforts to prevent the implementing of the CHT Accord are an indication of growing political activism among Bengali Muslims. One key informant argued that "the rights of Bengali settlers to form a political movement as well as to enjoy equal rights and privileges" (Expert 7). This is also the argument of the Bengali political groups in the CHT when they claim equal rights with the tribal people living in the CHT.

Indeed, it is difficult to determine the level of discrimination due to the absence of data. Underpinning the discrimination against tribal people is the lack of official recognition of their identity, the lack of protection of their rights by the police and legal system, and the lack of freedom of movement in their land (Mohsin, 1997, 2003; Chakma, 2010a). As the region is highly militarized, people cannot move freely. There is constant vigilance of the security forces to ensure the law and order situation in the CHT (D'Costa, 2016). Furthermore, high visibility of the security forces and structural dominance by the Bengali people at the top level of the local administration, such as police and bureaucracy, have created a sense of discrimination among the tribal people. Although the security forces are deployed in the CHT, they are not ensuring peace and law and order, and the laws are not protecting people's rights. Furthermore, the main security forces are now directly involved in business activities (tourism business) which go against the interests of the tribal people. The tribal respondents perceive the presence of the army as a policy of discrimination in the CHT as the security forces not only consume land but also complicate the social and economic life of the tribal people. In some respects, the army is involved in the practices, such as setting up the tourist spots (e.g., Nilgiri Resort, Sajek Valley resort) and seizing the land on the hills without informed consent and prior information (IWGIA, 2012; Cultural Survival, 2017). In every project, the army prefers to employ Bengali people to operate the activity, and this process helps to ensure the interest of the Bengali settlers and army as well. IWGIA (2012) notes that the army's conduct in the CHT is biased and discriminatory in ways that "often sympathise with, and more importantly, act in favour of the settlers, and against the indigenous people" (IWGIA, 2012: 12).

Marginalization

Another growing concern for the Bengali migration and settlement to the CHT is the marginalization of the tribal people in their land. Marginalization as a term refers to the exclusion of some people from the social and political system. It involves making some groups of people less important and placing them in a secondary position. In the CHT, marginalization is defined as the systematic disempowerment of the tribal people. This is also called the elimination, genocide, annihilation, or extinction of a community from their land (Levene, 1999; Adnan, 2004; Chakma, 2010b). This chapter calls it marginalization as the focus here is on the spatial marginalization of the tribal people in their own land. This marginalization is the result of the Bengali migration and settlement as well as the policy of the government. In the environment and conflict analysis, Homer-Dixon (2010: 77–78) describes "ecological marginalization as the marginal position of a group people due to the resource capture by the influential people in the face of the resource scarcity." A key informant shared that "Due to the encroachment of the Bengali settlers in the CHT, some tribal people have moved into the deep forest areas for their livelihood and to avoid conflict" (Expert 13). However, there are no statistics on how many people have moved as a result of being unable to live peacefully in their previous place. In some instances, Bengali people create an environment that the tribal people do not like. Expert 14 who worked in the remote area of the CHT offered the following comment:

> The activities of the Bengali people act as an encroachment into tribal society. The expansion of the religious institutions (for example, Mosques and Madrasas) is also pushing some tribal people to move long distances, possibly to the jungle so that they might have an environment for living.
>
> (Expert 14)

The mobilization of Muslim people and their eventual majority in the CHT has contributed to changing the basic demographic structure, means of economic activities, and political institutions. Now the tribal people are struggling to maintain their cultural distinctiveness due to the hegemonic nature of the Bangladesh state and intervention by the elite ruling classes (Arens, 2017; CHT Commission, 2000; Uddin, 2010). Even the traditional name of some villages and neighbourhoods has been changed due to the population having become majority Muslim. These places have taken on a Muslim cultural flavour and are known by Muslim names (Nasreen & Togawa, 2002; Nasreen, 2018).

Most of the key informants in the tribal background shared that due to the implementation of development projects, marginalized people, mostly of tribal origin, have become victims of eviction and displacement. Apart from the Kaptai Dam, other development projects, including a public university, medical college, eco-park, and tourist spots, have led to the marginalization of the people in the CHT. Experts of tribal background considered all these development projects as fostering tribal marginalization in the CHT. In response to a question concerning this, an expert suggested that "the public and medical university would open the door to many non-tribal students to come and study in the CHT, which would increase the dominance of the Bengali culture and practices" (Expert 6). In contrast, the informants in Bengali settlers argued that government presents these development activities in the CHT as part of the modernization of the region and to provide modern educational facilities to the tribal people as most of them struggle to come to cities for education. One key informant argued that "as they live in the remote area of Bangladesh, such development projects help them to foster economic development and get education facilities" (Expert 8). However, tribal key informants consider the development efforts to be part of the process of marginalizing tribal people. This suggests that the association of development with marginalization indicates a history of tribal people failing to reap the benefits of even well-meaning government projects.

This section has shown that respondents feel discriminated against by the state system. The tribal respondents show more disappointment about their life in the CHT. This disappointment has developed over the time of settlement of the Bengali people in their land. The Bengali respondents also feel discriminated against when they do not get equal treatment as citizens of the CHT with regard to what tribal people are getting from the state, namely, quota in the job sectors and admission to the educational institutions. The most crucial issue that has developed over time is the polarization between the communities. There is a sharp difference between the communities, and they have yet to come closer and cooperate in their daily life in the CHT. Many respondents do not want people from the other community as neighbours, do not trust them, and even fail to maintain a basic social relationship. The migration of the Bengali people has also influenced the social and economic condition in the CHT, which has marginalized poor tribal people through activities such as land grabbing and resource capture and forced some tribal people to move to the deep jungle for living and livelihood.

Fear, insecurity, and experience of violence

The CHT is marked as the most conflict-prone area in Bangladesh. The post migration situation in the CHT has characterized as conflictual and violent. The conflict situation generates fear and insecurity among residents in the CHT. Tribal residents feel fear and anxiety about land grabbing, the presence of the army, and the slow implementation of the Peace Accord (Chakma & D'Costa, 2013). The Bengali settlers are concerned about the competition over land and fear that the government undertaking to withdraw army camps from many locations of the CHT will make them more vulnerable to attacks. Two decades after the Peace Accord, the continued presence of fear and insecurity suggest an absence of peace, if peace is defined as the absence of violence as well as the condition of equality, equal rights, justice, free movement, and reciprocity among the people living in the CHT (de Rivera, 2018; Galtung, 1996).

Fear and insecurity

Although a sense of fear and insecurity is widely felt by the residents, it is likely to affect the Bengali and tribal communities differently. In order to explore these conditions, key informants were asked how often they feel fear and insecurity. A majority of the key informants argued that fear is not an everyday experience in their life although a majority of them feel fear frequently or sometimes. Tribal informants are more likely to feel fearful than their Bengali counterparts.

An informant from the tribal community mentioned that "they were very fearful of personal injury and attacks on their property, concern about their children's education, security of their family members, intimidation and verbal abuse" (Expert 4). The conflict situation both pre and post Peace Accord has implications for children's education. A study suggests that conflict situation in the CHT has prevented parents from sending their children to school (Badiuzzaman & Murshed, 2015). Moreover, people living in remote areas of the forest are often unable to send their children to school for security and financial reasons. In contrast, very few Bengali informants expressed heightened concern about these issues. Just over half said that they were "sometimes" concerned about them while a third of the Bengali respondents were not concerned. An expert mentioned that "Bengali settlers were very concerned for the extortion and their land" (Expert 8). In the CHT, the level of extortion is very high. The Bengali people also fear from eviction from their land.

Sources of insecurity

Fear and insecurity are the constant part of the CHT society. The post migration period has also marked with fear, insecurity, and violence in the CHT. The key informant and experts were asked about the issue of fear and insecurity in the CHT. Most of the key informants shared that they frequently (at least monthly) faced incidents of violence, while a significantly lower proportion of Bengali key informant were in this category. The key informants belonging to the Bengali community stated that they had not faced any violence or conflict in the last five years.

However, the tribal key informants were concerned about the physical attacks on themselves and/or a family member, and a similar proportion had their property destroyed. But Bengali respondents were less prone to falling victim to violence, with damage to property being the most likely event. However, the Bengali people were more likely to experience extortion than tribal respondents. The wide spread of a large number of illegal small arms and light weapons in the hands of outlaw people may cause such situations in the CHT. Jamil and Panday (2008) suggested that some of the members of the Shaniti Bahini did not surrender their arms to the Bangladesh government, and this group of people may be involved in collecting money from the general public. Although the respondents have termed this "extortion," Braithwaite and D'Costa (2012: 23) termed it "as 'tax' which is mandatorily given by the poor rural people to the members of PCJSS and UPDF for per unit of production."

Braithwaite and D'Costa (2018: 323) explain the current dimension of fear, sexual violence, organized crime, insecurity, and violence as the latest manifestation of violence gripping the region. This refers to the fact that the CHT has experienced violence, deprivation, and domination under different regimes from the British period to the current Bangladesh era. Although the Peace Accord was signed to manage the conflict and establish peace, the region is still undergoing violence, fear, and insecurities. This situation is far from being a sustainable peace. Thus, CHT experts have referred to the current situation of the CHT as an elusive peace (Panday & Jamil, 2009: 1052) and violent peace (D'Costa, 2016: 252). The CHT experts argued in this way because the absence of direct war between the army and tribal armed groups does not ensure a peaceful situation. According to the peace and conflict literature, this situation can also be referred to as negative peace (Galtung, 1969: 183). In order to achieve positive peace, some issues such as the presence of social justice, freedom, human rights, and choice of development are important preconditions which are only achievable with the elimination of structural violence from the society (Galtung, 1969: 183). In this sense, the CHT is still in a conflict situation.

The current state of conflict and actors

The previous section has analyzed the state of fear, insecurity, and violence in the CHT. The perceptions of the respondents about the fear, insecurity, and violence in their daily life give a picture of the conflict situation. There are different forms of "spoilers" such as military, organized groups, and government policy, which have resuscitated the conflict situation. The conflict situation arising from the data is a more small-scale, localized, and latent form of conflict that manifests with human rights violations, fear, insecurity, and land grabbing. The following section presents the current conflict situation and major actors in the conflict.

Different perceptions about peace and conflict

There is a significant difference between the tribal and Bengali respondents about the feeling of peace and conflict in the CHT. Key informants in the tribal community mention that they are not living in peace, whereas key informants in the Bengali community see themselves as living in a peaceful situation. To explain these responses, it is important to remember that the militarization, dominance, intergroup and intragroup conflict, and gender-based violence present the testimony of a conflict situation in the CHT (Braithwaite & D'Costa, 2018). More than 20 years after the peace treaty, many people still feel fear and insecurity. A recent study conducted by the Department of Peace and Conflict Studies, University of Dhaka, Bangladesh with the support of the UNDP-CHTDF Bangladesh comprised a household survey and an elite survey with 1,200 people as well as focus group discussions in the three districts. The research found significant indicators of conflict in the CHT such as displacement, communal violence, arms, drugs, extortion, polarization, and suspicion about the security forces (Choudhury et al., 2017). The study identified extortion and communal hatred as the most concerning issues that frequently cause fear and insecurity among the tribal and Bengali people living in the CHT.

The interviews conducted with an expert for the present study have also elicited information about the current state of conflict in the CHT. A conversation with the director of an NGO offers some insight into how the conflict has changed since the CHT Accord:

Researcher: Do you think that the peace accord has ended the conflict?

Expert 7: No, conflict is still ongoing. Even after the peace accord, many big forms of conflict have occurred in the CHT. I have visited some conflict spots and seen causalities. I saw that conflict is extensive in some places.

Researcher: Who is involved in this conflict?

Expert 7: Now, the conflict between the military and the Shanti Bahini has ended. However, the conflict between the tribal and the Bengali people is still ongoing, and innocent people are the victims.

Researcher: Can you tell me some incidents of conflict?
Expert 7: The major forms of the conflict is burning houses, destroying the crops of the opposing group, kidnapping people and demanding ransom, physically hurting people, even killing.

The informant sees ordinary people as the main protagonists of current civil strife in the CHT. This suggests that the nature of the conflict has changed from a violent war fought by armed forces to a small-scale social conflict (see the definition of conflict in the introduction chapter) taking place at the everyday local level. This social conflict takes place between the communities (Bengali settlers and tribal people) over resources, social position, and political power. This form of conflict is no less destructive as it has dire consequences for local people. The director of a civil society organization in the CHT, for example, mentioned that in Rangamati district a private medical hospital owned by the tribal community was destroyed during community violence (Expert 6). This raises the question of why the businesses, services, and crops owned by tribal people are often targeted in violent incidents. According to expert 6, "the motive of destroying the business centres by the Bengali settlers is to weaken the financial power of the tribal community."

Violence to small-scale conflict

In the post Peace Accord period, the conflict in the CHT can best be described as small scale and social conflict which is organized within the ethnic group along political divides (JSS vs. UPDF) or between the main ethnic groups (tribal vs. Bengali settlers). The manifestation of this activity is varied, including communal conflict, riot, civil strife, burning, killing, rape, and human rights violation (Mohsin, 2003). These incidents of conflict are more frequent in the daily lives of the CHT people, although some areas are more and others are less conflicted. The UN Special Rapporteur's report in 2011 outlines some gross human rights violations occurring in the CHT in the post Peace Accord period, such as arbitrary arrest, torture, extrajudicial killings, harassment, and sexual violence (cited in Chakma & D'Costa, 2013). This indicates that state and organizations are also involved in the conflict because ordinary people would not arbitrarily arrest people.

This raises questions about the perpetrators. While many acts of violence are carried out by ordinary people, according to the literature, there are other players who are directly or indirectly involved in originating and escalating conflict. Currently, in the CHT multiple actors are involved in addition to Bengali settlers and tribal residents. The fractional groups within the tribal people such as UPDF, PCJSS, UPDF reformist[7], and JSS reformist group are also involved. At the same time, some of them (e.g., UPDF) are directly opposing the Bengali settlers and the Peace Accord. The army also has a role in the conflict situation. Bengali and tribal respondents provided an idea about the actors in the current CHT conflict and violence.

A Bengali key informant argued that conflict is ongoing between settlers and tribal people (Expert 5). Tribal people are also involved in the conflict within the community as well as with the army. While the majority of Bengali respondents attribute agency to tribal people who they perceive to be fighting with settlers, army, and other tribal people, very few agreed with the idea that conflict exists among Bengali people (Expert 11). Conversely, another expert argued that Bengali settlers emerge as the main actors in the conflict. Bengali settlers were involved in the conflict with tribal people (Expert 13). A large majority also viewed Bengali settlers as being involved in the intragroup conflict (Expert 14). The Bengali people also agree that intragroup conflict also takes place among tribal people (Expert 3). In fact, there is conflict between the army and the tribal people, even though some experts and some of the literature assert that the army is no longer involved in the conflict.

Indeed, conflict between Bengali settlers and tribal people is most evident to the respondents, particularly to those of tribal background. Many of them have witnessed or directly experienced violent acts, as shown in the previous section. The role of the army is still questionable as respondents in both groups think that the army is actively involved in conflict. Chapter 3 and some experts (for example, expert 2) have already mentioned that the army has supported Bengali settlement; thus, conflict is less likely between the army and Bengali people. Experts in tribal background perceived that the army is involved in the conflict process in the CHT. The Bengali informants also see the army involved in the conflict with tribal people. While the army is not officially engaged in fighting, some members of the army may be engaged in small-scale conflict incidents such as rape, torture, and human rights violations. Reporting on violent incidents, D'Costa (2016: 248) identified "the involvement of the army in the CHT as the source of 'culture of violence as they use protracted and intense force in the region." D'Costa (2016, 254) also claimed that the army was involved in torturing tribal people and sexually abusing tribal girls and women on different occasions.

Intragroup conflict

A new phenomenon in the post Peace Accord era is the rise of intra-group conflict, a concerning issue for security and peace in the region. The intragroup conflict refers to the conflict among tribal people and Bengali people. From the interview with experts, it is perceived that conflict was happening mostly in the opposing group (i.e., Bengali settlers vs. tribal people). This indicates the absence of intragroup conflict within the own community, or, alternatively, the presence of intragroup conflict has been overstated in public discourse. Therefore, information from the experts and secondary sources may help to substantiate the arguments of intragroup conflict. Expert 5 compares conflict "within the community" and "conflict with the past rival." By conflict within the community, expert 5 refers to the conflict originating among the tribal people for the political power and social positions in the

CHT, and conflict over supporting the Peace Accord. It appears that tribal groups are sharply divided over the Peace Accord and engaged in sporadic violence and conflict. The competition for resources and political position, lack of consensus between and among groups, and opinion pro–Peace Accord and anti–Peace Accord are dividing tribal people in the CHT. It is also alleged that members of the contending groups, such as UPDF and PCJSS, are involved in killing tribal people who disagree with their political position (Panday & Jamil, 2009: 1065). For example, Panday and Jamil (2009: 1066) noted that

> the PCJSS and UPDF remain mired in an intra-communal conflict that has killed more than 500 people and injured about 1,000 since December 1998. Kidnapping and extortion by local gangs are frequent: more than 1,000 people have been kidnapped in the past 11 years.

However, Braithwaite and D'Costa (2012) give the responsibility for the conflict between the UPDF and PCJSS to the involvement of the law enforcing agencies (for example, DGFI). These researchers state that law enforcing agencies in the CHT are providing financial assistance and weapons to tribal fractional groups, such as UPDF, to fight against the PCJSS, which prolongs the battle between two fractional groups in the tribal community (Braithwaite & D'Costa, 2012: 21). This raises the question of why the law enforcing agencies are assisting one fractional group against another. The law enforcing agency may like to see the continuation of this battle with a view to weakening the collective force of the tribal people.

In the pre Peace Accord period, the major parties in conflict were the Bangladesh army and Shaniti Bahini, but in the post Peace Accord period, new players have emerged from the civilian people of both communities. On the indigenous side, tribal people in the CHT and other indigenous people living on the plains have formed a political platform named the Bangladesh Indigenous Peoples Forum (BIPF) to ensure the rights of the indigenous people in Bangladesh (Gerharz, 2014). Similarly, the Bengali settlers have formed a platform, Somo Adhikar Andolon (SAA), to ensure the rights and dignity of the Bengali people in the CHT. These two platforms seem to be rivals because each opposes the claim and position of others, and in addition, fractional political wings within the tribal community have emerged such as the UPDF and PCJSS reformist (D'Costa, 2014). Currently, the UPDF and PCJSS are the two main parties playing a role in the tribal community. On the other hand, SAA has emerged as a platform from the Bengali settlers to ensure their rights in the CHT. The army is still present, but their role has been officially transformed into ensuring security in the region. The PCJSS is a political party in the CHT with a long history of working for the rights and dignity of the hill people. The UPDF was formed in 1998 in the wake of the CHT Accord by PCJSS members who opposed the Accord and demanded complete freedom for the CHT.

Major players in the conflict

The experts think differently about the role of the military, Bengali settlers, UPDF, and PCJSS in the current conflict. Most experts with tribal background see the Bengali settlers and the army as the main actors in the conflict while the majority of the Bengali respondents consider the political tribal groups UPDF and PCJSS as the main actors of the conflict. This information suggests that each community has a different perception about the players of the current conflict situation. Multiple actors are influencing the social, economic, and political affairs in the region. For the multiple actors, the nature of conflict is complex and takes place in different forms, open or silently, and affects people to different extents. The goals of each actor are also different in nature. The UPDF and JSS, although local-based parties, actively participate in the national and local election and hold different processions and meetings for their ensuring their lights. The Bengali settlers are also divided along national political lines, such as BNP, Awami League, and Jamat-i-Islami. However, they hold the common view of ensuring their rights in the CHT.

The literature on CHT affairs confirms that the conflict situation has been complicated since the massive Bengali settlement and migration (Mohsin, 2003; Chakma, 2010a). The CHT remains the most militarized region in Bangladesh where security personnel are deployed to maintain peace and security. It is argued that the urge to protect the Bengali people in the CHT has necessitated the deployment of large numbers of military personnel in the region (Braithwaite & D'Costa, 2012; Adnan & Dastidar, 2011). The tribal people consider the security forces as one of the major sources of human rights violations of the tribal people. According to one study, there are 230 army camps in the CHT within a limited area of land (Mohsin & Hossain, 2015: 17), while another report puts the figure at around 35,000–40,000 security forces deployed in the region in 500 army camps (IWGIA, 2012: 12).

Despite army public relations programs to attain the confidence of the tribal people, such as the "pacification programme,"[8] the tribal people still consider the army as the cause of the conflicts and human rights violations. The role of the CHT political parties UPDF and PCJSS is also controversial, with some experts perceiving them to be responsible for creating conflict and violence. Some of the tribal background experts shared this view, perhaps because both parties are dominated by people from the Chakma ethnic group and perceived to marginalize other ethnic minorities. But more importantly, the PCJSS and UPDF are now rivals in every decision regarding the CHT, which is one of the important sources of generating violence, of the people of the Bengali and tribal communities (Braithwaite & D'Costa, 2012: 23). Other splinter groups have subsequently formed, and the emergence of new groups has been generating violence in tribal society.

This section, in summary, has argued that the CHT conflict is played out as a microlevel or social conflict between the tribal and Bengali people over the use of resources and social position. However, a new dimension to the conflict is the intragroup conflict between the UPDF and PCJSS people, which is a major source of killing and human rights violations. Due to the increasingly complex situation and their growing numbers, the Bengalis are also beginning to be politically organized through SAA, demanding equal rights in every sphere in the CHT. Now, the UPDF and SAA have fueled conflict and violence because they have radically different policies for the political future of the region. Above all, the role of the security forces (for example, the army) is still significant in implementing any decisions about peace and security in the region. At the same time, the presence of extensive security forces is hampering the human rights and individual freedom of the tribal people in the CHT.

Discussion on migration and conflict in the CHT

This section provides an overall discussion of the results presented in this chapter incorporating the theoretical explanations of climate change, environmental migration, and conflict. Bengali migration to the CHT has brought changes in the social, economic, and political context of the region which have influenced the environment, security, and conflict situation. Multidisciplinary approaches have been applied to analyze the relationship between the migration and settlement of the Bengali people and conflict situations in the CHT.

A significant change in Bengali migration has brought demographic transformation to the region. Demographic transformation and migration in an ethnically divided society sometimes generate conflict and violence. Countries with low economic capability, ethnic heterogeneity, and an illegitimate government are more prone to violence and conflict (Goldstone, 2002; Kahl, 2006; Urdal, 2005). Goldstone (2002) more precisely argues that population migration to an ethnically inhabited location changes the local balance, alters the dominance of the local people, and generates conflict and violence (Goldstone, 2002: 14). The sudden population increase caused by Bengali migration and settlement imposed tremendous pressure on resources and services and increased the level of violence in the CHT conflict during the period 1977–1997. Chakma and D'Costa claim that "two-thirds of the Bengali population in the CHT were migrants, and 62 per cent of the Bengali living in the CHT had been there for less than 30 years" (Chakma & D'Costa, 2013: 140). The massive migration during 1977–1989 and natural migration of the Bengali people due to their networks and connections have changed the unique characteristics and cultural patterns of the CHT and complicated the economic activities in the region. These changes in the culture, demography, and economic activities are acting to generate conflict between the tribal and Bengali communities. The demographic change has also resulted in resource scarcity that contributes to escalating the existing conflict and forming new kinds of conflict between the tribal and Bengali communities. The neo-Malthusian approach can best be used to analyze this situation of increasing population, resource scarcity, resource capture, and conflict (Homer Dixon, 2010; Kahl, 2006). In a poor and subsistence economic context, population pressure is a serious issue for security. Poor governance and discriminatory policies of the state sometimes complicate the situations for poor and marginal people. This is exactly what has happened in the case of the tribal people when a large number of environmentally induced migrants and poor people migrated to the region and attempted to share the resources in the CHT.

The migrant influx led to increased exploitation of land and forest resources, causing resource scarcity. In such conditions, the elites (political, civil, and military) have exploited the local people by seizing the land and forest resources. According to Homer-Dixon (1999: 4), "the scarcity of the critical resources – land and forest – generate severe social stress and unrest, and ultimately lead to conflict and violence." The local people opposed the Bengali migration and attacked settlements in some places. In response, the army and settlers burned the forest, destroyed and cleared it, and created a livable place for the Bengali people. The civil and military elites also took part in the process of resource capture and pushed many tribal people out of their land. As Homer-Dixon (2010: 7) points out, resource capture by the elites sometimes pushes people to be marginalized from their living options which can lead to conflict. In the CHT, the same situation has happened, and some of the civil and military bureaucrats stole land and plundered the forest resources, manipulating the state law (Roy, 2000; Adnan & Dastidar, 2011). As well as resource capture, land dispossession, deforestation, and development projects, the policy on the forest reserves and the expansion of mining operations have had significant impacts on the resource availability in the CHT.

The resource capture threatens the livelihood of the tribal people as they depended on natural resources for their livelihood. Currently, people of both the communities are very much concerned about their land, and land grabbing is now a leading cause of violence and human rights violation in the CHT. As land is the most important resource in the CHT, any form of disentanglement from the land pushes communities into disadvantaged and marginalized positions. It is argued that livelihood failure is connected with conflict, especially when the environmental causes contribute to the livelihood failure of a group of people (Ohlsson, 2000; Deligiannis, 2012). Land grabbing generates poverty by detaching some small communities from their land rights, leaving tribal people in poverty and struggling to receive proper help from the government (Jamil & Panday, 2008: 479). Migration, resource scarcity, and land grabbing generate a sense of deprivation among the tribal people which facilitates ethnic mobilisation.

Now, people in the CHT are highly polarized in ethnic, religious, and political points of view, which is contributing to forming social division, mistrust, and conflict. In any multiethnic society, inequality can lead to marginalization and discrimination along ethnic lines, generating ethnic hatred and social polarization (Østby, 2008a, 2008b; Stewart, 2010). The behaviour of the state and social faults caused by the Bengali migration have mobilized members of the tribal population to fight for their existence and livelihood. The survey results show a significant level of interethnic polarization, fear, insecurity, mistrust, and suspicion among both the Bengali and tribal people. It was clear that Bengali and tribal people do not like to have each other as neighbours, and they do not have a good relationship. As a consequence, both the Bengalis and tribal people are now polarized based on ethnic identity, which acts as a source of violence and conflicts in the region. The survey shows that people in both communities feel insecure, but the degree and sources of insecurity differ between tribal and Bengali respondents. Tribal respondents are more likely to feel insecure and make Bengali settlers and law enforcing agencies (army) responsible for it. Their feeling of insecurity is based on experiences of violence including verbal abuse, attacks on themselves or family members, attacks on their property, and loss of livelihood. Bengali respondents also report fear and insecurity due to their experiences of extortion and competition for land.

These negative attitudes hinder the communities from participating in social and political functions. Opportunities for sharing, neighbourhood activity, and participating in the joint social program help to enhance communication between and among the groups, which ultimately establishes peace and security in a multicultural society (Lake & Rothchild, 1996; Nagle, 2016; Oberschall, 2007). But in the CHT these opportunities are limited, and instead processes of discrimination and marginalization act to intensify the conflicting relationship between the Bengali and tribal people (Barkat, 2016).

The social and behavioural patterns of the tribal and Bengali settlers result in direct and indirect forms of conflict (Galtung, 1969, 1990). Direct violence includes killing, torture, and intimidation, and indirect violence involves deprivation, systemic subordination, marginalization, and systemic poverty (Galtung, 1990). According to survey findings, tribal people are more frequently exposed to direct violence. However, there is no official government information on how many people have been killed due to the conflict in the post Peace Accord period. For this reason, the information recorded by the IWGIA and PCJSS has been used to validate that conflicts still take place in the CHT, causing human rights violations, including the killing.

Conclusion

Migration and conflict interplay in Bangladesh exist because of the population pressure and resource scarcity. The inability of the governance system to balance between the increasing population and existing resources is also responsible for the ongoing conflict in Bangladesh. However, the conflict has a different story and background as it began with the denial of identity and with deprivation, destruction of property, and displacement of the tribal people. However, migration of the Bengali people has also contributed to these factors and has over time changed the focus, actors, and conduct of the conflict. Accelerated migration in the 1980 and 1990s increased the competition between the tribal and Bengali people over resources, most notably land, and also over social position and political power. Marginalization, discrimination, land grabbing, and human rights violations are now common in the CHT, creating widespread fear and insecurity among all ethnic communities and mutual distrust. In the urban slum, the migration and conflict interplay is mainly for the denial of basic needs of the people and human rights violation.

The state institutions have failed to develop more constructive relationships between and among the groups in the urban slums and in the CHT. People representing the local and national political parties are equally divided and hardly tolerate different opinions. The CHT conflict in its present form is more akin to a complex social conflict rather than an identity conflict. The divisive issues are multiple, including land, social position, development, immigration, and Islamization. Demographic pressure, changes in the ethnic composition, and resource exploitation have transformed the CHT into a more insecure place for both tribal and Bengali people. In particular, tribal people are threatened by the expansion of the neoliberal economy, exploitation of the resources by the capitalist Bengali influential people, and ongoing migration.

Additional migration in any form would further complicate the environment and may result in more conflict and violence in the slum and in the CHT. An appropriate policy to guide climate change adaptation and development of community relationships is essential to reduce the conflict between the tribal and Bengali people in the CHT. In the slum, management of internal migration is essential to ensure migrants' rights and curb the violence and conflict.

Notes

1 Kabuliyat is the official document against the land rights in Bangladesh. This document is given by land office situated in the local (*Thana/upozila*) level in Bangladesh to the citizen for the ownership of land. Chowdhury (2012) has referred it as the formal title of land ownership (Chowdhury 2012: 37).
2 In the CHT, common property is the slopes of the hills that are used by the tribal people without any formal document.
3 Islamization is widely used in the international arena to refer to a society transforming into a society upholding the Islamic culture, values, and principles in all spheres of life. In the CHT, complete Islamization is not possible; the society has deep attachments to tribal cultures and values. However, some researchers consider that the increasing number of mosques, madrassas, and renaming places with an Islamic name has started the process of Islamization in the 100 percent tribal society.
4 *Waz-Mahfil* is an Islamic lecture given by a religiously educated man which is organized by a group of people once a year. Different groups of people may arrange such occasions each year. This is popular in Bangladesh as a way to educate the general people about the Islamic values and culture. It is held at different

times and organized by some people who want to foster Islamic values in society. On this particular day of the year or religious calendar, one or several people are hired to give lectures in front of the people.

5 Somo Adhikar Andolon is the organization of the Bengali settlers in the CHT formed by Abdul Wadud Bhuiyan, a former MP of the Bangladesh Nationalist Party (BNP) from Khagrachari constituency. The supporters of this organization opposed the CHT Accord and demanded an equal share for Bengalis in all aspects of life with the tribal people in the CHT.

6 The Parbatya Bangali Chhatra Parishad was established in 1991 just after the Bangladesh Nationalist Party (BNP) formed the government. This student front works parallel to the Somo Adhikar Andolon in order to attain equal rights with the tribal people in the CHT in all sectors, including admission quotas for university places and jobs that currently exist for tribal students.

7 Reformists have emerged from both the PCJSS and UPDF group. In 1998, the UPDF emerged from the PCJSS due to the different opinions regarding the peace accord signed in 1997. The UPDF views the CHT problem differently, and thus, they perceive an alternative proposal for the rights of the CHT people. The Jano Samhati Samiti (Reformist) re-emerged as a political platform from 2007 and are working to increase their support base at the local level. Recently, the reformist group (Rupayan-Tanindra group) of the UPDF has emerged as a political platform. This friction among the tribal people is generating new forms of conflict and violence in the CHT.

8 The "pacification program" includes food distribution, construction of religious institutions or schools, small-scale income-generating projects, and so on. This program continued until the enacting of the Peace Accord.

References

Adnan, S. (2004). *Migration land alienation and ethnic conflict: Causes of poverty in the Chittagong Hill Tracts of Bangladesh*. Dhaka: Research & Advisory Services.

Adnan, S. (2008). Contestations regarding identity, nationalism and citizenship during the struggles of the indigenous peoples of the Chittagong Hill Tracts of Bangladesh. *International Review of Modern Sociology*, *34* (1), 27–45.

Adnan, S., & Dastidar, R. (2011). *Alienation of the lands of indigenous peoples: In the Chittagong Hill Tracts of Bangladesh*. Copenhagen: International Work Group for Indigenous Affairs. https://www.southasia.ox.ac.uk/sites/default/files/southasia/documents/media/oxford_university_csasp_-_work_in_progress_paper_15_adnan_dastidar_alienation_of.pdf

Ahmed, S., & Meenar, M. (2018). Just sustainability in the global South: A case study of the megacity of Dhaka. *Journal of Developing Societies*, 34(4), 401–424.

Amnesty International. (2013). *Pushed to the edge: Indigenous rights denied in Bangladesh's Chittagong Hill Tracts*. London: Amnesty International. https://www.amnesty.org/download/Documents/12000/asa130052013en.pdf

Arens, J. (2017). Genocide in the chittagong hill tracts, Bangladesh. In R Hitchcock (Ed.), *Genocide of indigenous peoples: A critical bibliographic review* (pp. 123–148). New York: Routledge. https://doi.org/10.4324/9780203790830

Badiuzzaman, M., Cameron, J. J., & Murshed, S. M. (2011). Household decision-making under threat of violence: A micro level study in the Chittagong Hill Tracts of Bangladesh. *MICROCON Research Working Paper No. 39*. Brighton: Institute of Development Studies, MICROCON. SSRN: https://ssrn.com/abstract=1762752, http://dx.doi.org/10.2139/ssrn.1762752

Badiuzzaman, M., & Murshed, S. M. (2015). Conflict and livelihood decisions in the Chittagong Hill Tracts of Bangladesh. In A. Heshmati, E. Maasoumi, & G. Wan (Eds.), *Poverty reduction policies and practices in developing Asia* (pp. 145–162). Asian Development Bank and Springer International Publishing AG.

Banks, N. (2008). A tale of two wards: Political participation and the urban poor in Dhaka city. *Environment and Urbanization*, *20*(2), 361–376.

Barkat, A. (2016). *Political economy of unpeopling of indigenous peoples: The case of Bangladesh*. Dhaka: MuktoBuddhi Prokasana.

Barkat, A., Halim, S., Poddar, A., Badiuzzaman, M., Osman, A., & ShanewazKhan, M. (2009). *Socio-economic baseline survey of Chittagong Hill Tracts*. Human Development Research Centre. Prepared for *CHTDF-UNDP, Dhaka*. https://www.yumpu.com/en/document/view/49581700/socio-economic-baseline-survey-of-chittagong-hill-tracts-chtdf

Barman, D.C. (2013). *Human rights report 2012 on indigenous peoples in Bangladesh*. Dhaka: Kapaeeng Foundation. http://www.kapaeeng.org/

Bianco, J. L. (2017). Resolving ethnolinguistic conflict in multi-ethnic societies. *Nature Human Behaviour 1*, 0085. https://doi.org/10.1038/s41562-017-0085 www.nature.com/nathumbehav

Braithwaite, J. B., & D'Costa, B. (2012). *Cascades of violence in Bangladesh. Peacebuilding Compared Working Paper*. Canberra: Australian National University. http://regnet.anu.edu.au/research/publications/2791/cascades-violence-bangladesh

Braithwaite, J., & D'Costa, B. (2018). *Cascades of violence: War, crime and peace-building across South Asia*. Acton, ACT: ANU Press, Peacebuilding compared. http://doi.org/10.22459/CV.02.2018

Chakma, A. (2017). The peacebuilding of the Chittagong Hill Tracts (CHT), Bangladesh: Donor-driven or demand-driven? *Asian Journal of Peacebuilding*, *5*(2), 223–242.

Chakma, B. (2010a). The post-colonial state and minorities: Ethnocide in the Chittagong Hill Tracts, Bangladesh. *Commonwealth & Comparative Politics*, *48*(3), 281–300. http://www.tandfonline.com/action/showCitFormats?doi=10.1080/14662043.2010.489746

Chakma, B. (2010b). Structural roots of violence in the Chittagong Hill Tracts. *Economic and Political Weekly*, *45*(12), 19–21.

Chakma, K., & D'Costa, B. (2013). The Chittagong Hill Tracts: Diminishing violence or violent peace, In E. Aspinall, R. Jeffrey, & A. Regan (Eds.), *Diminishing conflicts in Asia and the Pacific: Why some subside and others don't* (pp.137–49). Abingdon, Oxon: Routledge.

Chittagong Hill Tracts (CHT) Commission. (2000). *Life is not ours – land and human rights in the Chittagong Hill Tracts, Bangladesh*. https://www.iwgia.org/images/publications//0129_Life_is_not_ours_1-108.pdf

Chowdhury, R. (2012). *Land dispute resolution in the Chittagong Hill Tracts caught between liberalism and legal pluralism*. Masters thesis, McGill University. http://digitool.library.mcgill.ca/webclient/StreamGate?folder_id=0&dvs=1535617827479~938

Choudhury, Z. A., & Hussain, M. (2017). Dynamics of the CHT conflict. In Z. A. Choudhury (Eds.), *Conflict mapping in the Chittagong Hill Tracts (CHT) in Bangladesh* (pp. 127–145). Dhaka: Adhaorso Publication.

Choudhury, Z. A., Islam, R., & Alam, S. (2017). Micro-foundation of conflict in the CHT. In Z. A. Choudhury (Eds.), *Conflict mapping in the Chittagong Hill Tracts (CHT) in Bangladesh* (pp. 52–126). Dhaka: Adhaorso Publication.

Chowdhury, M. S. & Chakma, P. (2016). *Human rights report 2016 on indigenous peoples in Bangladesh*. Dhaka: Kapaeeng Foundation.

Chowdhury, S. R., Bohara, A. K., & Horn, B. P. (2018). Balance of power, domestic violence, and health injuries: Evidence from demographic and health survey of Nepal. *World Development*, *102*, 18–29.

Christie, D. J. (2009). Reducing direct and structural violence: The human needs theory. *Peace and Conflict*, *3*(4), 315–332.

Cultural Survival. (2017). *Observations on the state of indigenous human rights in Bangladesh*. Prepared for the *30th Session of the United Nations Human Rights Council Universal Periodic Review*. Massachusetts Avenue, Cambridge. https://www.culturalsurvival.org/sites/default/files/UPRBangladesh2017FINAL.pdf

D'Costa, B. (2014). *Marginalization and impunity-violence against women and girls in the Chittagong Hill Tracts*. Copenhagen, Denmark: International Work Group for Indigenous Affairs. Retrieved from https://issuu.com/iwgia/docs/cht_violence_against_women_study_20.

D'Costa, B. (2016). Journeys through shadows: Gender justice in the Chittagong Hill Tracts. In H. Hossain & A. Mohsin (Eds.), *Of the nation born* (pp. 234–282). Delhi, India: Zubaan.

de Rivera, J. (2018). Themes for the celebration of global community. *Peace and Conflict: Journal of Peace Psychology*, *24*(2), 216.

Deligiannis, T. (2012). The evolution of environment-conflict research: Toward a livelihood framework. *Global Environmental Politics*, *12*(1), 78–100. https://doi.org/10.1162/GLEP_a_00098

Esteban, J., & Schneider, G. (2008). *Polarization and conflict: Theoretical and empirical issues*. London: Sage Publications.

Feldman, S., & Geisler, C. (2012). Land expropriation and displacement in Bangladesh. *The Journal of Peasant Studies*, *39*(3–4), 971–993.

Fisher, R. J. (2016). Towards a social-psychological model of intergroup conflict. In *Ronald J. Fisher: A North American pioneer in interactive conflict resolution*. Pioneers in arts, humanities, science, engineering, practice series (Vol. 14, pp. 73–86). Cham, Switzerland: Springer.

Forsberg, E. (2008). Polarization and ethnic conflict in a widened strategic setting. *Journal of Peace Research*, *45*(2), 283–300. https://doi.org/10.1177%2F0022343307087185

Galtung, J. (1969). Violence, peace, and peace research. *Journal of Peace Research*, *6*(3), 167–191.

Galtung, J. (1990). Cultural violence. *Journal of Peace Research*, *27*(3), 291–305.

Galtung, J. (1996). *Peace by peaceful means: Peace and conflict, development and civilization*. Oslo: International Peace Research Institute.

Gerharz, E. (2014). Indigenous activism in Bangladesh: Translocal spaces and shifting constellations of belonging. *Asian Ethnicity*, *15*(4), 552–570.

Goldstone, J. A. (2002). Population and security: How demographic change can lead to violent conflict. *Journal of International Affairs*, *56*(1), 3–21. http://hdl.handle.net/1783.1/75219

Gurr, T. R. (1993). Why minorities rebel: A global analysis of communal mobilization and conflict since 1945. *International Political Science Review*, *14*(2), 161–201.

Gurr, T. R. (1995). *Minorities at risk- a global view of ethnopolitical conflicts*. Washington, DC: United States Institute of Peace Press.

Gurr, T. R. (2000). *Peoples versus states: Minorities at risk in the new century*. Washington, DC: United States Institute of Peace Press.

Gurr, T. R., & Moore, W. H. (1997). Ethnopolitical rebellion: A cross-sectional analysis of the 1980s with risk assessments for the 1990s. *American Journal of Political Science*, *41*(4), 1079–1103.

Halim, S., & Chowdhury, K. (2015). The land problem in the Chittagong Hill Tracts: A human rights anatomy. *JAMAKON Year Book* (pp. 36–80). Bangladesh: National Human Rights Commission. https://www.researchgate.net/publication/306322726_The_Land_Problem_in_the_Chittagong_Hill_Tracts_A_Human_Rights_Anatomy

Hasam, M. A., Arafin, S., Naznin, S., Mushahid, M., & Hossain, M. (2017). Informality, poverty and politics in urban Bangladesh: An empirical study of Dhaka City. *Journal of Economics and Sustainable Development*, *8*(14), 158–182.

Homer-Dixon, T. F. (1994). Environmental scarcities and violent conflict: Evidence from cases. *International Security*, *19*(1), 5–40.

Homer-Dixon, T. F. (1999). *Environment, scarcity, and violence*. Princeton, NJ: Princeton University Press.

Homer-Dixon, T. F. (2010). *Environment, scarcity, and violence*. Princeton, NJ: Princeton University Press.

Hossain, S. (2008). Rapid urban growth and poverty in Dhaka city. *Bangladesh E-journal of Sociology*, *5*(1), 1–24.

Hossain, S. (2013). Migration, urbanization and poverty in Dhaka, Bangladesh. *Journal of the Asiatic Society of Bangladesh (Humanities)*, *58*(2), 369–382.

International Work Group for Indigenous Affairs (IWGIA). (2012). *The indigenous world 2012*. Copenhagen, Denmark. https://www.iwgia.org/images/publications//0573_THE_INDIGENOUS_ORLD-2012_eb.pdf

Islam, M. R., & wa Mungai, N. (2016). Forced eviction in Bangladesh: A human rights issue. *International Social Work*, *59*(4), 494–507.

Jamil, I., & Panday, P. K. (2008). The elusive peace accord in the Chittagong Hill Tracts of Bangladesh and the plight of the indigenous people. *Commonwealth & Comparative Politics*, *46*(4), 464–489. https://doi.org/10.1080/14662040802461141

Jeong, H. W. (2005). *Peacebuilding in postconflict societies: Strategy and process*. Boulder, CO: Lynne Rienner.

Kahl, C. H. (2006). *States, scarcity, and civil strife in the developing world*. Princeton, NJ; Woodstock: Princeton University Press.

Kahl, C. H. (2018). *States, scarcity, and civil strife in the developing world*. Princeton, NJ: Princeton University Press.

Kalim, T., Hamadani, J. D., Kabir, R., & Majumader, M. A. A. (2017). Exploring the impact of intimate partner violence on children's behavior in urban slums of Dhaka City, Bangladesh, *Journal of Biostatistics and Epidemiology*, *3*(3/4), 95–105.

Kamruzzaman, M., & Hakim, M. A. (2015). Child criminalization at slum areas in Dhaka city. *American Journal of Psychology and Cognitive Science*, *1*(4), 107–111.

Khaleda, S., Mowla, Q. A., & Murayama, Y. (2017). Dhaka metropolitan area. In Y. Murayama, C. Kamusoko, A. Yamashita, & R. C. Estoque (Eds.), *Urban Development in Asia and Africa* (pp. 195–215). Singapore: Springer.

Khan, M. Z. S., Kundu, N., & Yeasmin, N. K. (2021). Sustainable quality of life of urban slum dwellers in Bangladesh: Evidence from Dhaka City. *Chiang Mai University Journal of Economics*, *25*(1), 62–71.

Koly, K. N., Islam, M., Reidpath, D. D., Saba, J., Shafique, S., Chowdhury, M., & Begum, F. (2021). Health-related quality of life among rural-urban migrants living in dhaka slums: A cross-sectional survey in Bangladesh. *International Journal of Environmental Research and Public Health*, *18*(19), 10507.

Lake, D. A., & Rothchild, D. (1996). Containing fear: The origins and management of ethnic conflict. *International Security*, *21*(2), 41–75.

Lata, L. N. (2017). Migration and urban Livelihoods: A translocal perspective in Dhaka, Bangladesh. In *Conference Proceedings TASA 2017 Conference*.

Lata, L. N. (2020). Neoliberal urbanity and the right to housing of the urban poor in Dhaka, Bangladesh. *Environment and Urbanization ASIA*, *11*(2), 218–230.

Latif, M. B., Irin, A., & Ferdaus, J. (2016). Socio-economic and health status of slum dwellers of the Kalyanpur slum in Dhaka city. *Bangladesh Journal of Scientific Research*, *29*(1), 73–83.

Levene, M. (1999). The Chittagong Hill Tracts: A case study in the political economy of creeping genocide. *Third World Quarterly*, *20*(2), 339–369.

Mohsin, A. (1997). *The politics of nationalism: The case of the Chittagong Hill Tracts, Bangladesh*. Dhaka: University Press Limited (UPL).

Mohsin, A. (2003). *The Chittagong Hill Tracts, Bangladesh: On the difficult road to peace*. Boulder, CO: Lynne Rienner.

Mohsin, A., & Hossain, D. (2015). Conflict and partition: Chittagong Hill Tracts, Bangladesh. In R. Manchanda (Ed.), *Sage Series in human rights audits of peace processes*, Vol. 5 (pp. 128–140). New Delhi, India; Thousand Oaks, CA: SAGE Publications.

Nagle, J. (2016). *Social movements in violently divided societies: Constructing conflict and peacebuilding*. Routledge Advances in Sociology series. London: Routledge.

Nahiduzzaman, K. (2006). *Housing the urban poor: Planning, business and politics: A case study of Duaripara Slum, Dhaka city, Bangladesh*. Master's thesis, Geografisk Institutt.

Nasreen, Z. (2018, June 4). Chittagong Hill tracts: The contested relationships and dynamics of politics, *Bangla Tribune* (English version). http://en.banglatribune.com/opinion/opinion/4101/CTG-Hill-tracts-The-contested-relationships-and

Nasreen, Z., & Togawa, M. (2002). Politics of development: 'Pahari-Bengali' discourse in the Chittagong Hill Tracts. *Journal of International Development and Cooperation*, *9*(1), 97–112.

Northrup, T. A. (1989). The dynamic of identity in personal and social conflict. In L. Kriesberg, T. A. Northrup, & S. J. Thorson (Eds.), *Intractable conflicts and their transformation* (pp. 55–82). Syracuse University Press.

Nurul, S., & Mohammad, I. (2014). State of water governance in Dhaka Metropolitan city of Bangladesh: Evidence from three selected slums. *International Journal of Interdisciplinary and Multidisciplinary Studies*, *1*(2), 19–38.

Oberschall, A. (2007). *Conflict and peace building in divided societies: Responses to ethnic violence*. London: Routledge.

Ohlsson, L. (2000). *Livelihood conflicts: Linking poverty and environment as causes of conflict*. Swedish International Development Cooperation Agency (SIDA). https://www.sida.se/contentassets/99c24545bf31484aa0e6bbd7658a5873/livelihood-conflicts-linking-poverty-and-environment-as-causes-of-conflict_1326.pdf

Østby, G. (2008a) Inequalities, the political environment and civil conflict: Evidence from 55 developing countries. In F. Stewart (Ed.), *Horizontal inequalities and conflict: Understanding group violence in multiethnic societies* (pp. 136–159). Basingstoke: Palgrave Macmillan.

Østby, G. (2008b). Polarization, horizontal inequalities and violent civil conflict. *Journal of Peace Research*, *45*(2), 143–162. https://doi.org/10.1177%2F0022343307087169

Panday, P. K., & Jamil, I. (2009). Conflict in the Chittagong Hill Tracts of Bangladesh: An unimplemented accord and continued violence. *Asian Survey*, *49*(6), 1052–1070.

Paul, B. K. (2006). Fear of eviction: The case of slum and squatter dwellers in Dhaka, Bangladesh. *Urban Geography*, *27*(6), 567–574.

Paul, B. K., & Ramekar, A. (2018). Internal migration in Bangladesh: A comparative analysis of coastal, environmentally challenged, and other districts. In R. McLeman & F. Gemenne (Eds.), *Routledge handbook of environmental displacement and migration* (pp. 225–237). Routledge.

Peacock, A. C. (Ed.). (2017). Introduction: Comparative perspective of Islamization. In *Islamization: Comparative perspectives from history*. Edinburgh: Edinburgh University Press.

Rahman, M., Jahir, T., Yeasmin, F., Begum, F., Mobashara, M., Hossain, K., Khan, R., Hossain, R., Nizame, F. A., Leontsini, E., Unicomn, L., Luby, S. P., & Winch, P. J. (2021). The lived experiences of community health workers serving in a large-scale water, sanitation, and hygiene intervention trial in rural Bangladesh. *International Journal of Environmental Research and Public Health*, *18*(7), 3389.

Ramsbotham, O. (2005). The analysis of protracted social conflict: A tribute to Edward Azar. *Review of International Studies*, *31*(1), 109–126.

Reuveny, R. (2007). Climate change-induced migration and violent conflict. *Political Geography*, *26*(6), 656–673. https://doi.org/10.1016/j.polgeo.2007.05.001

Reuveny, R. (2008). Ecomigration and violent conflict: Case studies and public policy implications. *Human Ecology*, *36*(1), 1–13. https://doi.org/10.1007/s10745-007-9142-5

Roy, R. C. K. (2000). *Land rights of the indigenous peoples of the Chittagong Hill Tracts, Bangladesh*. IWGIA Document No. 99. Copenhagen: International Work Group for Indigenous Affairs. https://www.iwgia.org/images/publications//0128_Chittagong_hill_tracts.pdf

Roy, R. D. (2002). *Land and forest rights in the Chittagong Hill Tracts*. Kathmandu, Nepal: International Centre for Integrated Mountain Development (ICIMD).

Saha, S. (2012) Security implications of climate refugees in urban slums: A case study from Dhaka, Bangladesh. In J. Scheffran, M. Brzoska, H. Brauch, P. Link, & J. Schilling (Eds.), *Climate change, human security and violent conflict* (Vol. 8, pp. 595–611). Hexagon Series on Human and Environmental Security and Peace. Berlin, Heidelberg: Springer.

Stewart, F. (2010). *Horizontal inequalities and conflict: Understanding group violence in multiethnic societies*. Conflict, Inequality and Ethnicity Series. Basingstoke, Hampshire: Palgrave Macmillan.

Tajfel, H., & Turner, J. C. (1979). An integrative theory of intergroup conflict. *The Social Psychology of Intergroup Relations*, *33*(47), 74.

Taras, R., & Ganguly, R. (2015). *Understanding ethnic conflict*. London: Routledge.

Thapa, G. B., & Rasul, G. (2006). Implications of changing national policies on land use in the Chittagong Hill Tracts of Bangladesh. *Journal of Environmental Management*, *81*(4), 441–453. https://doi.org/10.1016/j.jenvman.2005.12.002

Uddin, N. (2010). Politics of cultural difference: Identity and marginality in the Chittagong hill tracts of Bangladesh. *South Asian Survey*, *17*(2), 283–294.

Ullah, M. S., Shamsuddoha, M., & Shahjahan, M. (2014). The viability of the Chittagong Hill Tracts as a destination for climate-displaced communities in Bangladesh. In S. Leckie (Ed.), *Land solutions for climate displacement* (pp. 215–247). London: Routledge.

Urdal, H. (2005). People vs. Malthus: Population pressure, environmental degradation, and armed conflict revisited. *Journal of Peace Research*, *42*(4), 417–434.

7 Conclusion

Introduction

This book has investigated the relationship between climate change and migration; and between migration and conflict in Bangladesh. More specifically, this book has a new explanation for the CHT conflict and urban areas. This connection between climate change–induced migration and conflict is unexplored in Bangladesh even though the country is one of the most climate affected countries, and climate affected people have already migrated to the CHT and urban areas. The migration to the urban areas, particularly in Dhaka, is already addressed in the academic and research arena, but in the context of CHT this exploration is yet to exist. As such, it is a nascent topic as well as a new perspective on the CHT conflict analysis. Existing analysis of the CHT conflict is predominantly centered on the conventional study of hegemony, denial of the identity of the tribal people by the ruling classes, militarization, and gross human rights violation by many actors, including the security forces (Mohsin, 1997; Levene, 1999; D'Costa, 2012). This book has provided alternative explanations by investigating the Bengali migration and conflict in the CHT and following their links with environment and climate change issues. The issue of civil unrest, violence, and conflict in the urban center is also explained in the line of political and economic perspective. This book has given analysis on the basis of climate change, migration, and conflict theory. Broadly, three fundamental issues guide this research: the role of climate change as a driver of Bengali migration to the CHT and migration of the climate affected people to the urban areas; the examination of the processes and mediating factors which institutionalize migration in the urban centers and the CHT; and the influence of migration in the CHT conflict formation. Connecting these three issues gives the book a unique importance in the area of climate change, migration, and conflict analysis in a climate hot spot.

For connecting the three separate issues – climate change, migration, and conflict – existing literature has been reviewed and a gap identified in the context of Bangladesh as well as the CHT. This book has filled up the identified research gap around the role of climate change and produced new

DOI: 10.4324/9781003430629-7

information on how climate change events have led to the Bengali migration and influenced the conflict in the CHT. In the context of the CHT conflict, some existing research (mostly international literature) identified that landlessness and environmental factors have contributed to the migration to the region (Lee, 1997, 2001; Reuveny, 2007). However, empirical research to explore the connection has been lacking. Moreover, the dominant interpretation in Bangladesh (Mohsin, 1997; Adnan, 2004; Adnan & Dastidar, 2011) presents this migration as a state-sponsored intervention into the CHT conflict. Specifically, it is argued that successive governments hoped that Bengali settlement would defuse the conflict by weakening local Jumma identity and political claims for autonomy. This book utilized the climate change and migration literature as an analytical lens to conduct a ground level study in the CHT and found that climate change events and poverty influenced the decisions of the Bengali people to migrate to the CHT, with consequences for the CHT conflict. This explanation of climate change, migration, and conflict relationship reflects the theoretical framework (Figure 3.1), which implies that environmental and climatic events along with economic and social issues paved the way of migration and settlement of the Bengali people and generated the conflict situation in the region. This offers an alternative explanation of the CHT conflict which goes beyond mainstream perspectives of "political migration or "state-sponsored settlement."

While analyzing climate change, migration, and conflict relationships, this book has also explained the migration of the poor and climate induced migrants to the urban areas, particularly in Dhaka city, which is expanding the city horizontally and changing the structure of the city. The increasing population in the cities is the outcome of climate change and displacement of the poor people from their land. Climate change and migration in urban areas are widely discussed in Bangladesh (Kartiki, 2011; Ahsan, Karuppannan, & Kellett, 2011; Ahsan, Kellett, & Karuppannan, 2016; Banks, Roy, & Hulme, 2011; Hassani-Mahmooei & Parris, 2012; Herrmann & Svarin, 2009). However, the climate change–induced migration and conflict relationship is unexplored in the context of urban areas. This book has presented this complex connection of climate change, migration, and conflict in the urban areas in Bangladesh.

Although migration is an ongoing process and commonly used to avoid vulnerabilities as well as pursue a better life (McLeman, 2014), climate change–affected rural people in Bangladesh do not like to migrate. They are attached to their land, and many do not have the resources to migrate (Gray & Mueller, 2012). Sometimes, climate affected people migrate only a short distance, hoping to return to their place of origin after the vulnerabilities subside (Lein, 2000). Nevertheless, on many occasions climate affected people migrate permanently with the help of a number of mediating factors ranging from state to nonstate actors. In reviewing the literature on factors mediating migration, this book has explored that government support, security forces, and networks played a prominent role in the case of Bengali migration to the CHT. Bur the migration of the people to the urban areas is a natural process which is mainly occurring for livelihood and shelter. Sometimes, social connections help migrants to come and settle in the slum.

There has also been limited research on how the migration of Bengali people has transformed the social, economic, and political fabric of the CHT, making it more complex and competitive. Traditional conflict analysis, incorporating the militarization, failure of the political leaders to give constitutional recognition to the tribal people, and human rights violations, is partial and incomplete. Migration, demographic transformation, and resource scarcity have the potential to escalate the existing conflict and generate new conflict in the CHT, a situation which is either missing or ignored in existing research. This neglect has impacted the analysis of conflict and conflict management. By assessing and acknowledging climate change–induced migration and its impact on the CHT conflict, this thesis has the potential to inform more appropriate policies related to climate change adaptation and conflict management. By studying the internal migration process and identifying the multiple factors including climate change impacts on the lowlands, this research contributes to a better understanding of the multiple interconnected measures required to better manage climate change migration.

The empirical data from the key informants and experts have provided insights to develop three findings chapters: a new perspective on climate change–induced internal migration in the urban areas and the CHT; the mediating factors that contribute to the migration of the poor and climate affected people; and how migration and conflict interplay in the urban areas and the CHT in Bangladesh. The theoretical framework, derived from blending the two strands of literature – climate change and migration; climate change–induced migration and conflict – as well as the contextual study of Bangladesh have informed each chapter. These chapters are interconnected and provide new insight into the connections between climate change, migration and conflict. The following section provides a summary of the research findings.

Climate change and migration interplay

Climatic events have profound impacts on the displacement and migration of people from their place of origin to urban areas and the CHT. Bengali people migrated to the CHT in response to climatic events and poverty in their place of origin. Climatic events including floods, riverbank and coastal erosion, and drought contributed to their migration decision through destroying their livelihood options and leaving them in extreme poverty. Interviews with key informants and experts helped to explain how climatic events and poverty are interwoven in lowland and rural areas in Bangladesh, confirming existing research (Ahmed, Diffenbaugh, & Hertel, 2009; Hertel, Burke, & Lobell, 2010). People in coastal regions are the most affected, not least because these regions are impoverished, and households have few resources to cope with adverse climate events (World Bank, 2016).

Furthermore, climatic events are so frequent and so severe in some places in Bangladesh that they displace many people from their homes. Information from the key informants confirm theadverse effects of climatic events on their homes, agricultural land, and livelihood options. Having to frequently face severe climatic events made living impossible in their place of origin. Climatic events predominantly cause poverty through crop failure, shrinking the livelihood options, and damaging or destroying living spaces. Yet most people did not immediately move to another place for shelter and livelihood, but instead tried to remain in, or return to, their original homes. Only when climate affected people felt deep dissatisfaction about the place where they lived and saw no prospect of improvement did they decide to migrate in order to escape hardship and poverty. The possibility of getting land and the availability of forest resources in the CHT worked as a pull factor, encouraging climate affected and poor people to migrate to the region. Equally, climate-affected people migrate to the urban areas for getting scope to work in readymade garment industries and other nonformal industries. Some people also migrate seasonally for pulling rickshaws in Dhaka, which is a source of income for the poor people. The uneven development in Bangladesh and lack of decentralization of industries are also the causes for the migration of the poor and climate-affected people in the urban areas.

Although climate and environmental change displaces many people in Bangladesh, it does not necessarily push people to migrate to a particular place. Like other forms of migration (both internal and international), the migration process depends on multiples causes, situations, connections, and the possibility of a more favourable environment. This is the case in the Bengali migration to the CHT, where a number of mediating factors assisted the migration process. One of the main contributions of this book is to explore the mediating factors based on the interview responses, showing that the resettlement policy of the government, protection from the security forces, and information and informal assistance through the migrants' networks encouraged the Bengali people to migrate to the CHT. Extending the current understanding of the Bengali settlement as a political measure to weaken the Jumma identity movement and defuse the CHT conflict (Adnan & Dastidar, 2011), Chapter 5 argues that the migration process was mediated by several factors, including civil administration, the army, social networks, and family connections. Thus, the climate-induced migrants and poor people were able to actualize their expectation of settling in the CHT and urban areas. Without these mediating factors, migration of the poor and climate-affected people would not have been possible in the conflict region. In the CHT, ethnic minority people not only opposed the Bengali migration but also were engaged in violent conflict with the army and settlers to protect their special status and recognition of their right to self-determination.

Climate change and conflict interplay

This book has unveiled the relationship between climate change–induced migration and conflict interplay in Bangladesh. This relationship is not linear but a complex relationship in which multiple issues act as triggering factors. A literature review and expert interviews have highlighted the current conflict situation in the urban areas and in the CHT. They argue that the social, economic, and political conditions of both the places become complicated by the population pressure and increasing migration. The condition of the CHT has become increasingly contested as a result of large-scale migration and settlement from the 1970s onward. Migration has significantly contributed to a three-fold increase in the population of the region between 1974 and 2011 and has transformed the CHT into a place where intense pressure on existing resources and land is felt by tribal inhabitants, and increasingly also by Bengali inhabitants. Tribal people have resisted the Bengali settlement policy, and in the post Peace Accord period leaders have urged the Bangladesh government to stop issuing residency certificates to Bengali people in the CHT (Jamil and Panday 2008).

If the settlement policy was meant to be a climate and environmental change adaptation policy and a poverty alleviation policy, it has not worked well, particularly for the reluctant host community. The government wants to ensure healthy living in urban areas, particularly in Dhaka. However, the migration of the people forces the government to compromise with the services that are provided to the urban dwellers. The scarcity of the resources, poor governance system, and corruption act as the triggering factors to generate conflict and violence in the urban areas. In the case of the CHT, a competition for land and resources has increased between the migrant and local tribal people. The local tribal people who depend on the community forest and *jhum* cultivation are struggling to make a living and are pushed to more marginal locations in search of land and livelihood, as well as to escape the encroachment of the Bengali people. Not only Bengali settlers but also some civil and military bureaucrats have been engaged in land grabbing, which has become one of the main sources of conflict and violence in the area.

Currently, the CHT is in the grip of social conflict that catches media attention when it is manifested through violence, including killing, burning houses, destroying property, rape, attack, and extortion. However, this study has shown that the social conflict is underpinned by deep mistrust between the Bengali and tribal communities, whose views on issues are often highly polarized. Polarization and mistrust have formed over time due to the long-standing conflict and are continuously reinforced by experiences of land grabbing, human rights violation, and double standards in the ways in which the civil and military institutions in the CHT treat people. Depending on their ethnicity and status, people feel insecure, for example experiencing threats to life and physical health, abduction, and extortion. Although a high military presence continues in the region, ostensibly to maintain peace and security, people from both communities feel insecure and threatened. The sources of the threats and insecurities are known to them, but for various reasons, people do not seek justice from the law enforcement agencies. This is due to a lack of trust in the law enforcing agencies, which have neglected conflict situations or have been a source of insecurity and involved in committing crimes against the local tribal people. For example, on many occasions tribal people claimed that members of the law enforcement abused tribal girls and women (Guhathakurta, 2012; Mohsin, 2013; D'Costa, 2014). The migrant Bengali people also feel insecure and experience threats from armed gangs and outlaw groups. This suggests that the security system has failed to protect people from various ethnic groups, but mistrust between the Bengali and tribal people also acts as one of the drivers of insecurity and violence in the CHT.

The Dhaka city and the CHT have continued to witness frequent events of small-scale and violent conflict. Though not a regular phenomenon, these events frequently affect the lives of the people. Threats, rape of a family member, torture, damage to houses and other property, and extortion are some common forms of direct violence in the urban areas and the CHT society. But the roots of the violence in the urban areas are manifold. In the case of the CHT, direct violence immediately affects people, while indirect violence such as discrimination, lack of facilities, and marginalization of people (Galtung, 1969) prevents society from moving forward and attaining security, justice, and development. In many cases, minor incidents become the source of significant conflict. The pre-existing disputes or criminal incidents such as sexual abuse, killing, abduction, and altercations become triggers for violent conflict that violates the human rights of the people.

Contributions to theory

This book contributes to the theoretical, policy, and practical levels of understanding about climate change, migration, and conflict. Research into when and how climate change–induced migration impacts conflict situations, or creates new conflict, has been hampered by a lack of empirical and local level case studies (Gleditsch, 2012; Hendrix, 2018; McLeman, 2014). Most of the research has focused on the relationship between climate change and migration, or on the relationship between climate change and conflict. This book contributes to advancing understanding of climate change and migration, and the climate change and conflict relationship by teasing out the links between these three related issues in the context of Bangladesh – one of the most climate affected countries in the world. It has drawn on Homer-Dixon's (1999) analysis of resource scarcity and violence and Kahl's (2006) analysis of demographic pressure and civil strife, which highlight the role of migration, population pressure, and resource scarcity in the emergence of violent civil conflict. While researchers have acknowledged that conflict may originate when people with different ethnic backgrounds migrate to an area (Goldstone, 2002; Kahl, 2006), most do not explain how conflict may occur in such situations. Moreover, some researchers argued that climate change may accelerate conflict in such places where there is already a conflict situation. Climate change effects thus exacerbate an existing conflict. From this perspective, the current book is an important case study–based document as there was already conflict in the CHT when large-scale migration occurred, and within this conflict situation, the migration has aggravated the conflict. This causal relationship between Bengali migration and conflict was ignored and unexplored in the body of research on the CHT conflict.

This book, therefore, contributes to the theoretical understanding by adding insights into the notion that climate change–induced internal migration of people with different backgrounds may endanger the cultural distinctiveness and existence of minority people in other parts of the world. This book advances the idea that the failure of the government to address the needs of climate change displaced people, and at the same time, the government's assistance in settling displaced people in a minority-dominated location, have had crucial impacts on the conflict and insecurity in the region. The human insecurity issues of indigenous people due to increasing climate change impacts on their lives and livelihoods has been acknowledged by international research (Adger et al., 2014). Although this research identifies the effects of climate change on the indigenous people, it does not consider the human security situation of the indigenous people as and when the climate displaced people with different backgrounds from other parts migrate to the indigenous inhabited areas. The book has explored this issue and provided evidence that climate change–induced internal migration with people from different backgrounds causes conflict and violence in the indigenous inhabited locations, as it creates pressure on sharing existing resources (e.g., land and forest).

By advancing the concept of climate change–induced migration and conflict in the CHT, this book offers alternative explanations of the CHT conflict which are largely absent in the national and international literature. In doing so, this book opens up avenues for the development of policies both for addressing climate change–induced internal migration and the CHT conflict. A better understanding of the relationship between climate change migration and conflict in the CHT would also assist researchers in other countries who are working in conflict and peace studies where the environment and climate change is the nascent topic. The state policy for managing the population flow in the urban areas is also a crying need for ensuring the rights of the climate migrants as well as maintaining the livability and sustainability of the urban areas.

Policy implication and practice

This book has strong policy implications. As this book explored the implications of climatic events in the migration and migration and settlement of the affected people to the CHT and urban areas, it sought also to provide recommendations that move beyond the traditional analysis of the conflict and security studies. It is argued that the CHT conflict has strong connections with the migration and settlement of Bengali people which has transformed the demographic structure of the region. This can be described as "displacing conflict"[1] (Swain, 1996) which refers to environmentally displaced people migrating to another place where they eventually contribute to conflict. In the CHT, a large number of Bengali poor and climate change induced migrants moved to the tribal inhabited area and complicated the existing conflict.

Thus, this book advances two issues: First, a need for more control and management of internal migrants by appropriate policies by the government and donor organizations so that displaced people could survive in their place of origin; and second, a more proactive policy to build the capacity of people to cope with climate change events. Like other parts of Bangladesh, the CHT has a scarcity of arable land as well as land suitable for dwellings. Moreover, in-migration of Bengali people has a strong connection with generating conflicts and violence in the region; therefore, further in-migration of Bengali people will put pressure on the land and existing resources as they will compete with the tribal people (Ullah et al., 2014). Instead of encouraging migration to places already involved in conflict, governments should focus on building the resilience capacity of people in lowland and highly climate affected areas. If this was done, poor and displaced people would not need to migrate to the CHT, which should reduce migration-related conflicts.

This book supports the arguments for including more detailed analysis of internal migration in climate policies.[2] In particular, climate policymakers need to consider more seriously the possibilities of internal instability and conflict arising from using migration as climate change adaptation. In the current action plans and policies adopted by the government of Bangladesh, the research and knowledge management section only mentions that "climate change-related internal and external migration and rehabilitation should be monitored" (MoEF, 2009: 58). The revised version (2009) of the Bangladesh Climate Change Strategy and Action Plan (BCCSAP) gave emphasis to addressing the migration issue either by discouraging it or by providing resettlement options. Internal migration and policy issues are missing in the action plan because of the ignorance of internal migration issues. Moreover, migration of the Bengali people to the CHT is neither recognized nor documented. The action plan also lacks guidelines on how conflicts that arise from climate change–induced migration should be managed.

This book is written based on in-depth study to explore the relationship between climate change–induced migration and conflict in Bangladesh. Thus, this book has the contribution to encourage policymakers to rethink the connection between climate change and migration to the urban areas and the CHT, as well as the implications of migration to local conflict. More research is needed, including research in other parts of Bangladesh, especially in cities, to investigate how climate change–induced migration may be a factor in social conflict. Since most climate change affected people migrate to urban locations, such a study would help to explore the link between climate change–induced migration and conflict. Second, the policy must be directed toward managing the present conflict, minimizing the insecurities, bridging gaps between communities, and building sustainable peace in the CHT. Although multifaceted programs are underway by joint ventures between the government and international organizations since the signing of the Peace Accord, the Bengali migration issue is overlooked. Moreover, reconciliation between Bengali settlers and tribal people is a prime policy concern to manage the current conflict.

This book also offers information that could guide practical measures to manage the social conflict between the Bengali and tribal communities in the CHT as well as managing the civil conflict and violence in the urban areas. In the CHT, people and political groups of both communities are engaged in conflict and violence for resources, social, and political positions. Law enforcement agencies are also involved to some extent in generating violence when they violate the rights of the tribal people. To address this situation, both communities and state agencies must be brought under a single peacebuilding umbrella in order to build trust, cooperation, and reconciliation. In a post-conflict or a conflict situation, peacebuilding measures such as demilitarization, democracy, security sector reform, ensuring human rights, and providing basic needs work as a stimulus to bridge the gaps and help communities to cooperate for the shared interests and values (Jeong, 2005; Lambourne, 2000).

In the CHT, peacebuilding processes are ongoing in order to bridge the gap between the communities; however the local ownership of the resources, land, and social positions are not included in these initiatives. In a post-conflict situation, local ownership of the resources and involvement in the decision-making process constitute crucial factors in building lasting peace (Donais, 2012). Migration and settlement of the Bengali people and land grabbing have contested the ownership of land, resources, and social positions. Despite the efforts of the government and NGOs, the region still experiences communal violence and conflict. Bengali migration, along with other issues, prolongs the conflict and delays the peacebuilding in the region. Many of the principles of the Accord have yet to be implemented, and issues such as demilitarization, security sector reform, and ensuring human rights are key responsibilities the state still needs to fulfil. Resolving land disputes and creating an effective land commission may help to defuse the social conflicts between tribal people and Bengali settlers. Thus, it is an important step to implement the Peace Accord as was promised by the government of Bangladesh.

The land grabbing by multinational corporations and powerful people must be controlled to ensure the survival of the poor communities. The commercial use of land, rubber plantations, and tourism businesses help the outsiders, not the community people who depend on the land for their existence. This encroachment by the powerful elites and corporations is now regarded as "contemporary colonialism" which is acting against the indigenous people (Alfred & Corntassel, 2005) and necessitates dispossession and disconnection of the local people from their land and resources. It is clear that land is being rapidly taken by the powerful people and corporations, which is alarming for both the tribal and Bengali people; however, tribal people are the worst victims of this land grabbing and land acquisition. Tourism and conservation, if necessary, must be operated ensuring the benefit of the local people; otherwise in the future more resentment and conflict will be generated.

Notes

1 Swain refers to Bengalis who have been displaced due to environmental destruction and livelihood failure in Bangladesh and migrated to Assam and West Bengal, where they have become a source of conflict.
2 Climate policy means the action plan and polices that the Bangladesh government has already adopted for the adaptation and migration with increasing climate change, for example NAPA, BCCSA, fifth five-year plan.

References

Adger, W. N., J. M. Pulhin, J. Barnett, G. D. Dabelko, G. K. Hovelsrud, M. Levy, Ú. Oswald Spring, & C. H. Vogel. (2014). Human security. In C. B. Field, V. R. Barros, D. J. Dokken, K. J. Mach, M. D. Mastrandrea, T. E. Bilir, M. Chatterjee, K. L. Ebi, Y. O. Estrada, R. C. Genova, B. Girma, E. S. Kissel, A. N. Levy, S. MacCracken, P. R. Mastrandrea, & L. L. White (Eds.), *Climate change 2014: Impacts, adaptation, and vulnerability* (pp. 755–791). *Part A: Global and Sectoral Aspects. Contribution of Working Group II to the Fifth Assessment Report of the Intergovernmental Panel on Climate Change.* Cambridge and New York: Cambridge University Press.

Adnan, S. (2004). *Migration land alienation and ethnic conflict: causes of poverty in the Chittagong Hill Tracts of Bangladesh.* Dhaka: Research & Advisory Services.

Adnan, S., & Dastidar, R. (2011). *Alienation of the lands of indigenous peoples: In the Chittagong Hill Tracts of Bangladesh.* Copenhagen: International Work Group for Indigenous Affairs. https://www.southasia.ox.ac.uk/sites/default/files/southasia/documents/media/oxford_university_csasp_-_work_in_progress_paper_15_adnan_dastidar_alienation_of.pdf

Ahmed, S. A., Diffenbaugh, N. S., & Hertel, T. W. (2009). Climate volatility deepens poverty vulnerability in developing countries. *Environmental Research Letters*, *4*(3), 034004.

Ahsan, R., Karuppannan, S., & Kellett, J. (2011). Climate migration and urban planning system: A study of Bangladesh. *Environmental Justice*, *4*(3), 163–170.

Ahsan, R., Kellett, J., & Karuppannan, S. (2016). Climate migration and urban changes in Bangladesh. In R. Shaw, A. Rahman, A. Surjan, G. Parvin (Eds.), *Urban disasters and resilience in Asia* (pp. 293–316). Amsterdam: Butterworth-Heinemann.

Alfred, T., & Corntassel, J. (2005). Being indigenous: Resurgences against contemporary colonialism. *Government and Opposition*, *40*(4), 597–614.

Banks, N., Roy, M., & Hulme, D. (2011). Neglecting the urban poor in Bangladesh: Research, policy and action in the context of climate change. *Environment and Urbanization*, *23*(2), 487–502.

D'Costa, B. (2014). *Marginalisation and impunity-violence against women and girls in the Chittagong Hill Tracts*. Copenhagen, Denmark: International Work Group for Indigenous Affairs. https://issuu.com/iwgia/docs/cht_violence_against_women_study_20

D'Costa, B. (2012). *Nationbuilding, gender and war crimes in South Asia*. Singapore: Routledge.

Donais, T. (2012). *Peacebuilding and local ownership: Post-conflict consensus-building*. Singapore: Routledge.

Galtung, J. (1969). Violence, peace, and peace research. *Journal of Peace Research*, *6*(3), 167–191.

Gleditsch, N. P. (2012). Whither the weather? climate change and conflict. *Journal of Peace Research*, *49*(1), 3–9. https://doi.org/10.1177%2F0022343311431288

Goldstone, J. A. (2002). Population and security: How demographic change can lead to violent conflict. *Journal of International Affairs*, *56*(1), 3–21. http://hdl.handle.net/1783.1/75219

Gray, C. L., & Mueller, V. (2012). Natural disasters and population mobility in Bangladesh. *Proceedings of the National Academy of Sciences*, *109*(16), 6000–6005.

Guhathakurta, M. (2012). *The Chittagong Hill Tracts (CHT) Accord and after: Gendered dimensions of peace* (pp. 197–214). New Delhi: Routledge.

Hassani-Mahmooei, B., & Parris, B. W. (2012). Climate change and internal migration patterns in Bangladesh: An agent-based model. *Environment and Development Economics*, *17*(6), 763–780. https://doi.org/10.1017/S1355770X12000290

Hendrix, C. S. (2018). Searching for climate–conflict links. *Nature Climate Change*, *8*(3), 190–191. https://doi.org/10.1038/s41558-018-0083-3

Herrmann, M., & Svarin, D. (2009). Environmental pressures and rural-urban migration: The case of Bangladesh. Geneva: UNCTAD.

Hertel, T. W., Burke, M. B., & Lobell, D. B. (2010). The poverty implications of climate-induced crop yield changes by 2030. *Global Environmental Change*, *20*(4), 577–585.

Jamil, I., & Panday, P. K. (2008). The elusive peace accord in the Chittagong Hill Tracts of Bangladesh and the plight of the indigenous people. *Commonwealth & Comparative Politics*, *46*(4), 464–489. https://doi.org/10.1080/14662040802461141

Jeong, H. W. (2005). *Peacebuilding in postconflict societies: Strategy and process*. Boulder, CO: Lynne Rienner.

Kahl, C. H. (2006). *States, scarcity, and civil strife in the developing world*. Princeton, NJ; Woodstock: Princeton University Press.

Kartiki, K. (2011). Climate change and migration: a case study from rural Bangladesh. *Gender & Development*, *19*(1), 23–38. https://doi.org/10.1080/13552074.2011.554017

Lambourne, W. (2000). Post-conflict peacebuilding. *Security Dialogue*, *31*, 357.

Lee, S.-W. (1997). Not a one-time event: Environmental change, ethnic rivalry, and violent conflict in the Third World. *The Journal of Environment & Development*, *6*(4), 365–396.

Lein, H. (2000). Hazards and "forced" migration in Bangladesh. *Norsk Geografisk Tidsskrift*, *54*(3), 122–127. https://doi.org/10.1080/002919500423735

Levene, M. (1999). The Chittagong Hill Tracts: A case study in the political economy of creeping genocide. *Third World Quarterly*, *20*(2), 339–369.
McLeman, R. A. (2014). *Climate and human migration: Past experiences, future challenges*. New York: Cambridge University Press.
MoEF. (2009). *Bangladesh Climate Change Strategy and Action Plan 2009*. Ministry of Environment and Forests, Government of the People's Republic of Bangladesh, Dhaka, Bangladesh, xviii + 76pp.
Mohsin, A. (1997). *The politics of nationalism: The case of the Chittagong Hill Tracts, Bangladesh*. Dhaka: University Press Limited (UPL).
Mohsin, Z. R. (2013). The crisis of internally displaced persons (IDPs) in the federally administered tribal areas of Pakistan and their impact on Pashtun women. *Tigah: A Journal of Peace and Development*, *3*(2), 92–117.
Reuveny, R. (2007). Climate change-induced migration and violent conflict. *Political Geography*, *26*(6), 656–673. https://doi.org/10.1016/j.polgeo.2007.05.001
Swain, A. (1996). Displacing the conflict: Environmental destruction in Bangladesh and ethnic conflict in India. *Journal of Peace Research*, *33*(2), 189–204.
Ullah, M. S., Shamsuddoha, M., & Shahjahan, M. (2014). The viability of the Chittagong Hill Tracts as a destination for climate-displaced communities in Bangladesh. In S. Leckie (Ed.), *Land solutions for climate displacement* (pp. 215–247). London: Routledge.
World Bank Group. (2016). *World development report 2016: Digital dividends*. World Bank Publications.

Appendixes

Appendix 1: Introduction to the climate migrant informants

Respondent	*Gender*	*Background*
Climate migrant 1	M	A 45-year-old who migrated from Bhola district due to floods and sea-water intrusion and now runs a small business selling betel leafs and cigarettes in Naikonchari in Khagrachari district
Climate migrant 2	F	A 55-year-old who migrated from Khulna district due to floods and now lives in Khagrachari Sadar in Khagrachari district
Climate migrant 3	M	A man of about 50 years of age who migrated from the Chittagong district due to floods and cyclones and now lives in the Bengali Para in Khagrachari district
Climate migrant 4	M	A 50-year-old who migrated from Barishal district due to coastal erosion and now lives in Longudu
Climate migrant 5	F	A 50-year-old who migrated from Faridpur district due to riverbank erosion and now lives in Khagrachari
Climate migrant 6	M	A 45-year- old who migrated from Chittagong due to floods and now lives in Khagrachari Sadar
Climate migrant 7	M	A 50-year-old who migrated from Bagerhat district due to floods and coastal erosion and now lives in Bandarban district
Climate migrant 8	M	A 55-year-old who helped many people from Barishal district, where he also came from. Now he is the leader of his neighbourhood (*para*) in Bandarban district
Climate migrant 9	M	A 45-year-old who migrated from the Sylhet district due to floods and now lives in Rangamati district
Climate migrant 10	M	A 25-year-old who migrated from Chandpur district and now runs a business in the CHT. He leads a local chapter of the Bengali Student Association

(*Continued*)

Respondent	*Gender*	*Background*
Climate migrant 11	F	A 45-year-old who migrated from Khulna district due to floods and lives in Korail slum in Dhaka
Climate migrant 12	M	A 44-year-old who migrated from Khulna district due to floods and now lives in Korail slum in Dhaka
Climate migrant 13	M	A 50-year-old who migrated from the Chittagong district due to floods and cyclones and now living in a slum in Dhaka
Climate migrant 14	F	A 50-year-old who migrated from Patuakhali district and now live Dhaka slum
Climate migrant 15	N	A 40-year-old who migrated from Gaibandha district due to riverbank erosion and now lives in a slum in Dhaka
Climate migrant 16	F	A 45-year-old who migrated from Bhola due to cyclone and periodic floods and now lives in Korail slum
Climate migrant 17	M	A 50-year-old who migrated from Bagerhat district due to coastal erosion and now lives in a slum in Dhaka
Climate migrant 18	M	A 55-year-old who came from Barishal district and now lives in Korail slum in Dhaka
Climate migrant 19	M	A 45-year-old who migrated from the Sylhet district due to floods and now lives in a slum in Dhaka
Climate migrant 20	M	A 40-year-old man from Rangpur district who now lives in a slum in Dhaka

Appendix 2: Introduction to the experts

Respondent	*Gender*	*Profession*	*Expertise*
Expert 1	M	Chief executive of a research and development institute	Background in sociology, researches on climate change adaptation and mitigation issues
Expert 2	M	University professor	Expert in minority rights, social movements
Expert 3	F	University professor	CHT expert and think tank
Expert 4	F	Development NGO worker	Development worker in the CHT
Expert 5	M	Indigenous political leader	Expert in the CHT issues
Expert 6	M	Director of an NGO	Expert in migration, legal issues, and CHT conflict
Expert 7	M	President of national NGO	Experience in the fields of minorities, CHT, and environment issues

(*Continued*)

Respondent	*Gender*	*Profession*	*Expertise*
Expert 8	M	Senior research fellow at a national research institute	Expert in environment and climate change–induced security issues
Expert 9	F	University professor and director a local research organization	Expert in the field of climate change, migration, and adaptation issues
Expert 10	F	Chairperson of a research organization	Expert in minorities, CHT, women's rights, and migration issues
Expert 11	F	University professor	Expert in social forestry, climate change, and migration issues
Expert 12	M	Director of a peace research institute	Expert in security and conflict issues
Expert 13	F	University teacher and activist	Expert in anthropology and CHT affairs
Expert 14	F	Community leader	Expert in identity, politics, and CHT affairs
Expert 15	M	Independent researcher	Expert in environment and security issues

Index

Pages in *italics* refer to figures.

Printed in the United States
by Baker & Taylor Publisher Services